SATELLITE AND CABLE TV SCRAMBLING AND DESCRAMBLING

Brent Gale

Frank Baylin

Published by Baylin/Gale Productions
Cover Illustration by Peter Stollard

For speaking engagements or technical consultation contact:

Frank Baylin
1905 Mariposa
Boulder, Colorado 80302
(303) 449-4551

or

Brent Gale
4462 Driftwood Place
Boulder, Colorado 80301
(303) 530-3147

Artist: Amy Lockard

First Edition

First Printing: June 1986

ISBN: 0-917893-07-7
LCCCN: 86-070985

DEDICATION

We dedicate this book to all the cats and all the mice working so diligently to better each other and without whom the "cat and mouse" game would not be possible.

ACKNOWLEDGEMENTS

Many people contributed to helping us create this book. However, due to the sensitive nature of the material, they all desired to remain anonymous. We respect their wishes.

SATELLITE AND CABLE TV SCRAMBLING AND DESCRAMBLING

FOREWORD

Ever since the discovery of the Rosetta Stone, mankind has been fascinated with the techniques used to encrypt and decode data in various forms. Up until about 25 years ago, the general public's familiarity with cryptography was limited to accounts in the popular press about secret codes used by spies and detectives to convey covert messages. Recreational crptography was popularized by Sir Arthur Conan Doyle via Sherlock Holmes and eventually inspired syndicated newspaper features. Video encryption appeared publicly in the late 1950's with the first pay-TV station in Hartford, Connecticut. The consumer's attitude towards paying a fee for television programming was somewhat less than enthusiastic, and the service folded shortly after it began. When cable TV systems began to emerge in isolated areas, the public eagerly accepted the idea of paying for improved reception, the prime advantage of cable in the early days of its existence. A natural extension of this concept, i.e. "premium" programming, became a national reality when Time-Life began broadcasting Home Box Office over C-band satellites in the mid-1970's. To assure payment for this service, some method of denying program access to non-subscribers was required. Development of consumer video encryption schemes began in earnest. In the collective mind of the cable TV industry, at least part of the security of any encryption system lies in the denial of general access to information about the encoding process. On the other hand, the average consumer needs to be educated concerning the positive and necessary aspects of video encryption. Until now, technical information (and misinformation) regarding video encoding and decoding technology has been available to the consumer primarily through the "pirate underground." This book is not intended to foster piracy or the "cracking" of scrambling codes. What it does do is take the mystery out of video encryption, and educate the interested reader in a manner that is clear, concise, and eminently readable. It is a worthwhile addition to the reference library of any student of modern video technology.

Dr. Stephen J. Bepko
Baltimore, Maryland
June 1, 1986

I. INTRODUCTION

Our book is written for readers not necessarily having detailed technical educations who desire to have an understanding of the technology underlying satellite and cable television broadcasting and scrambling. However, we have attempted to provide as much useful information as possible knowing that this will serve all participants in our rapidly growing business by speeding its natural evolution.

Encryption systems have been developed solely to protect the economic rights of producers and marketers of television broadcasts. However, the fair and appropriate use of scrambling technologies can be as much a positive development for consumers as it is for programmers and program distributors. The list of beneficiaries is long and includes film makers, cable TV companies, satellite broadcasters, businesses and consumers in even the most remote geographic corners. The successful marketing of pay-per-view entertainment and information means not only that producers will be compensated but that new, innovative broadcast materials will proliferate.

The mechanics by which consumers have paid for video entertainment has evolved considerably since the pioneer days of television broadcasting. Conventional over-the-air television has been financed almost exclusively by commercials. By contrast, cable television networks have been supported by monthly subscription fees collected from each subscriber. The most primitive scrambling technique has been and, in some cases, still is employed. This protection is provided by simply connecting or disconnecting a cable at each customer's residence. However, over the years, cable engineers have been forced to develop more sophisticated scrambling and decoding equipment in order to protect their ownership rights and to maintain their economic viability in the face of talented pirates. The dynamic interaction between programmers and these creative do-it-yourselfers has inadvertently forced the creation of ever more secure and sophisticated encryption systems. And the process is accelerating especially with the advent of high powered, low cost microcomputers.

Until recently, satellite TV broadcasts have either been commercially sponsored or have been offered free-of-charge, either intentionally or unintentionally. No satisfactory encryption methods had been available to allow "pay-TV" producers to collect revenues. The development of scramblers and matched decoders which have evolved from and beyond those used on cable TV networks in combination with spin-offs from sophisticated military encryption systems has ushered in a new and revolutionary era in satellite television. Customers scattered anywhere over an entire continent and beyond can now be addressed and managed from one central control computer. The economic potential is staggering.

There is an intense debate surrounding the issue of the fair use of scrambling devices. The legality of using privately owned or built devices to descramble satellite and cable TV broadcasts is also currently a target of substantial litigation. In certain instances owners of "pirate" decoders have been successfully prosecuted by broadcasters who have used "theft of service" arguments. However, the controversy is not settled. Efforts to brand owners of home satellite TV receiving stations pirates failed in the United States with the passage of a bill in late 1984 that legalized reception of unscrambled satellite broadcasts. And there is now a widely supported "grass roots" movement afoot whose objective is to force programmers to make their scrambled shows available at a fair price and from competitive sources. Many feel that fair marketing means access via other than exclusively cable TV networks. Our book is not intended as a judgement of the legal and social issues.

Numerous groups are now marketing circuit diagrams or kits designed to be capable of descrambling television broadcasts. In some cases, completed "black boxes" are available for purchase. Hobbyists can certainly derive educational benefits from building and understanding the ins and outs of descrambling circuitry. However, programmers do deserve fair compensation for their efforts. And readers should also be aware that among the honest vendors of circuit diagrams are some unscrupulous "quick-buck" artists who feed on the anger and resentment generated by those pricing practices considered unfair by many consumers.

Our book is organized so that the material in each chapter builds upon that explored in previous ones. The concepts underlying modern communications are outlined in Chapter I. In the next chapter, the operation of both black/white and color televisions are comprehensively explained. This includes an examination of the differences between the various broadcast standards used throughout the world and the transmission of stereo sound along with television. Chapter IV is an in-depth study of the scrambling techniques which underlie all encryption systems in use today. The innovative MAC broadcast system is also outlined here.

In Chapters V and VI, the discussion turns practical as commercially available scramblers and descramblers for both cable and satellite TV systems are presented. This study includes an explanation of both the technical parameters of each system as well as its operational features.

We would appreciate feedback from our readers if there are any notable omissions or incorrect facts. Writing an accurate and up-to-date book of this nature is a challenge not without pitfalls, given how rapidly the underlying technologies are evolving.

II. UNDERLYING CONCEPTS

Some basic concepts are fundamental to understanding the material that follows on television and encryption systems. This information is included to make our book complete but an electrical engineer or PhD in physics can comfortably skip this chapter. However, for the majority of us not having such rigorous technical backgrounds, the objective of providing a clear and in-depth treatment based on a few fundamentals is certainly feasible.

A. ELECTROMAGNETIC WAVES

An electromagnetic wave is similar in concept to the waves that travel outward in concentric circles when a pebble is tossed into a pond or to the waves of vibrating air molecules that we know as sound. While sound travels at a plodding 1,223 kilometers per hour (760 miles per hour), radio and all other electromagnetic waves in free space travel at the speed of light, 299,339 kilometers per second (186,000 miles per second). At this rate, a signal completes the journey from an uplink antenna to a satellite and back to earth in about four tenths of a second, or from a television station or along a cable network to a customer virtually instantaneously.

Frequency

The frequency of an electromagnetic wave or electrical signal relayed by wire is the number of vibrations that occur every second as the energy passes any point in space. Just as the frequency of sound vibrations determines whether a musical note is either a soprano or a bass, so the frequency of radio waves determines whether they are used to transmit regular AM radio or satellite television broadcasts. Microwaves have frequencies in excess of one billion cycles per second

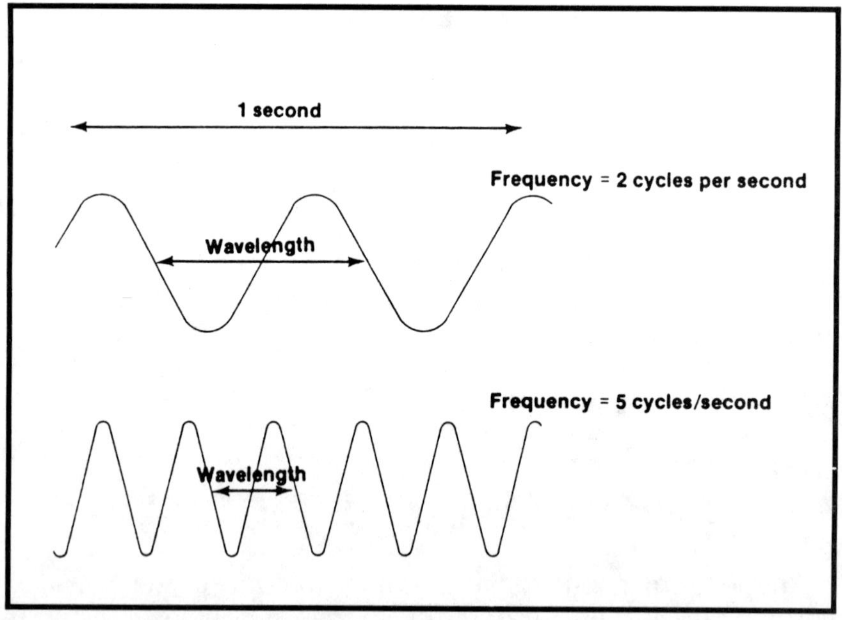

Figure 2-1. Wavelength and Frequency. *The two fundamental but related characteristics of electrical and electromagnetic waves are frequency and wavelength. Higher frequency signals have shorter wavelengths and vice versa.*

(one gigahertz, abbreviated at 1 GHz). By comparison, the line current from a wall outlet has a frequency of 60 cycles per second (60 Hertz) in North America and 50 Hertz in Europe. In other words, the voltage changes from positive to negative either 60 or 50 times each second.

Many seemingly different phenomena encountered in nature including light, X rays, infrared heat rays, microwaves used for both cooking and communicating, and gamma rays from the cosmos are electromagnetic waves. Surprisingly, the only difference among them is their frequency. Since all electromagnetic waves travel at the same speed, the speed of light, as the frequency increases the wavelength must decrease. For example, the wavelength of visible light is comparable to the dimensions of atoms and molecules, and their frequencies are many billions of billions of cycles per second. In contrast, microwaves have lower frequencies of one to fifty GHz and wavelengths ranging from 30 centimeters to a few millimeters. Radio waves have wavelengths which can be kilometers long and frequencies in the millions of cycles per second (megahertz or MHz) range.

TABLE 2-1. FREQUENCY ABBREVIATIONS

NAME	ABBREVIATION	MEANING
Hertz	Hz	Cyles per second
Kilohertz	KHz	Thousands of cycles/second
Megahertz	MHz	Millions of cycles/second
Gigahertz	GHz	Billions of cycles/second

Phase

The simplest form of electomagnetic or electrical wave is a pure sine wave. For example, line current is a 60 cycle sinewave. The phase of one sine wave is always measured relative to another reference wave or starting point. Not knowing the reference would be like timing a 100 yard dash without knowing when it began. One complete cycle is considered, like one complete rotation on a circle, to be 360 degrees.

So if the peaks of two waves are one quarter of a wavelength apart, they have a 90 degree phase difference.

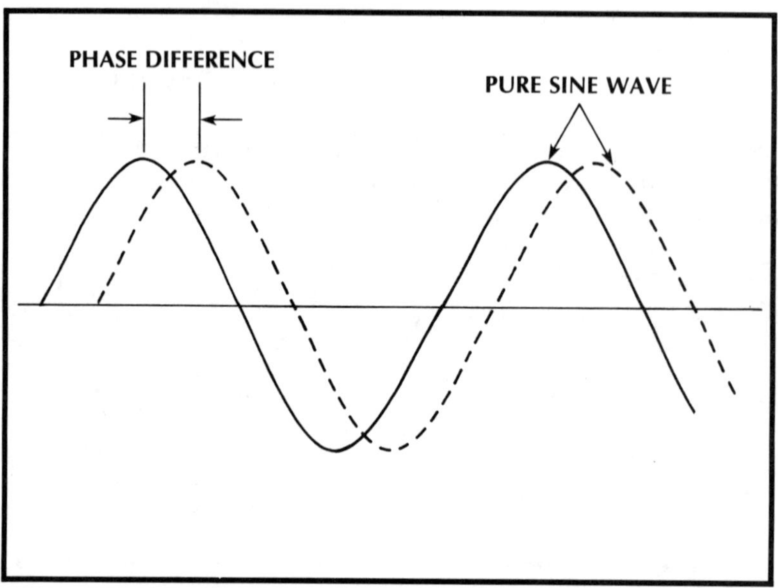

Figure 2-2. Phase. *Phase defines the relationship between two signals. The signals shown here are approximately one eighth of a wavelength or 45 degrees (one eighth of 360 degrees) out of phase with each other. One wave is said to lead and the other to lag by 45 degrees.*

Power

The power of electromagnetic waves is commonly measured in watts or watts per square meter. For example, 10 watts per square meter means that the power passing through each square meter is ten watts. Satellite TV broadcasts are usually received by an antenna at powers less than one billionth of one billionth of a watt per square meter!

The power of a signal relayed over a conductor is commonly measured in watts. Therefore, for example, an electronic circuit might transform a 1 milliwatt signal into a more powerful one of 10 watts. This amplification or gain uses external power to act upon and to boost the signal power.

B. COMMUNICATION FUNDAMENTALS

The same principles underlie all forms of man-made communication. The first step is creating and coding the message. Next this information must be modulated or added onto the medium designed to carry the signal. At the receiving end, the signal is demodulated and the original information is extracted. The amount of information carried is determined by its "bandwidth." The power of the signal can be amplified (increased) or attenuated (decreased). And unwanted signals or noise are always present to hinder perfect communication.

Coding the Message — Analog and Digital Signals

Any message, whether it be the image and voice of an entertainer or details of stock-market transactions, must first be changed into a form that can be relayed over-the-air by electromagnetic waves or on a conductor via electrical waves. Analog coding methods mimic the pattern of a message by changes in electrical voltages. For example, a voice can be transformed into an analog signal by a microphone that creates a voltage pattern determined by the loudness and frequency of the sound. The louder the sound, the higher the voltage. The higher the sound frequency, the more rapid changes occur in this voltage.

By contrast, a digital coding method uses only the numbers 0 and 1 to convey all information about these voltage levels and frequencies. For example, a voice could be expressed in digital form if the loudness at each point in time was expressed by numbers, or patterns of 0s and 1s. Or a photograph can also be described by a long series of 1s and 0s that are coded so that some impart information about the location of the dots composing the picture while others determine the brightness and color of the dots. Computers exclusively use digitally coded messages.

Communication devices can relay either digital or analog forms of the same message. Analog-to-digital (A/D) or digital-to-analog (D/A) converters can translate between these two forms of representation. For example, conversations between computers relayed by satellite are always digital, while most TV broadcasts are expressed in analog form. Compact disc players convert digital streams detected by a laser from the surface of a revolving plastic disc via a D/A converter into analog waveforms that can drive loudspeakers.

Modulation — Adding the Message to Carrier Waves

Analog or digital signals are impressed upon radio or microwaves by a process called "modulation." Once the message is modulated onto the carrier wave of an acceptable frequency, it can be relayed from a sending to a receiving location whether it is via a satellite link, over-the-air or along a cable route. The term carrier wave describes precisely what the modulated wave does, it carries the information. Radios, televisions and other communication equipment demodulate or extract the original message from the carrier wave. The demodulation process is also known as detection.

The simplest method to modulate a carrier wave is to switch it on and off. For example, Morse code can be relayed as a series of dots and dashes by turning the carrier wave on and off. The most familiar methods of modulation are amplitude modulation (AM) and frequency modulation (FM) as encountered, for example, in AM and FM radio broadcasts. In amplitude modulation, the power of a carrier wave is varied in accordance with the voltage level of the message being relayed, while in frequency modulation the frequency of the carrier wave is varied.

Each of these types of modulation has advantages and disadvantages. On one hand, AM messages must have relatively high powers to be capable of travelling long distances without being weakened too

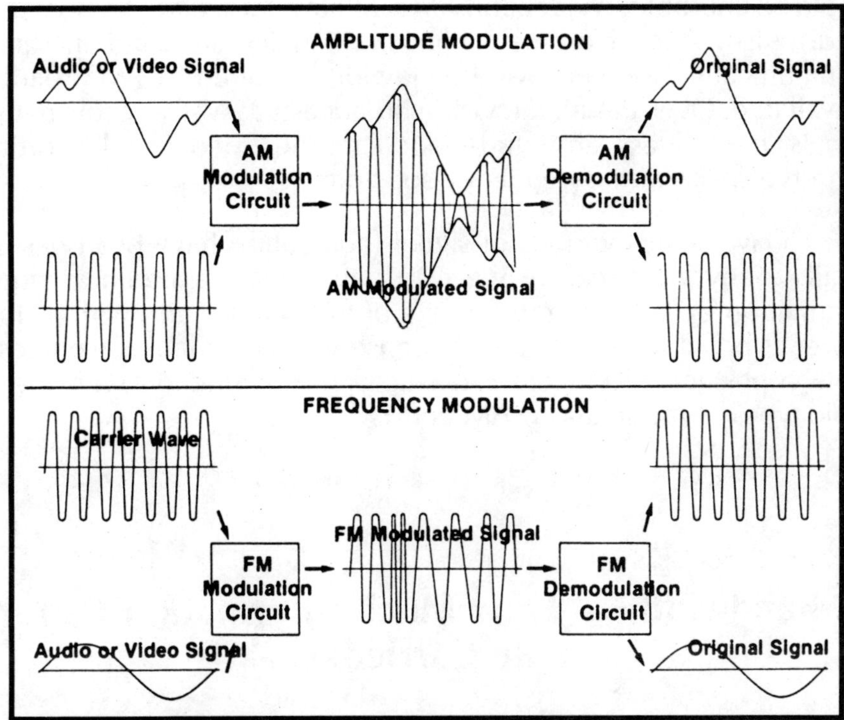

Figure 2-3. AM and FM Modulation. *Amplitude and frequency modulation are the basic techniques used to impress a signal onto a carrier wave. This modulated audio, video or data information can then be transmitted by cable, over-the-air or by satellite to its final destination. A demodulator then extracts and restores the original signal.*

severely by atmospheric disturbances and other forms of noise. They are also more prone to picking up static or interference than are FM messages. On the other hand, FM signals generally need relatively lower power for successful, long distance transmission, because they must use a substantially wider band of frequencies than AM messages do to carry the same amount of information.

Satellite transmissions are frequency modulated. As signals cover the long distances between uplinks, satellites and earth receiving stations, their power becomes so low that standard AM transmissions would be unusable. (Wideband AM satellite audio relays are sometimes used.) Also, since extremely high frequencies are used in satellite communication the very wide bandwidth required by FM broadcasts is available. Over-the-air television broadcasts as well as cable network signals, by contrast, amplitude modulate the video signal in order to conserve space in the frequency spectrum.

Today, other methods of signal modulation have been designed to allow any given carrier wave to transmit a maximum amount of information over as narrow a range of frequencies and with as low a power as possible. Such efforts to conserve the available resources are quite sophisticated but are often used in conjunction with the two basic types of modulation, AM and FM.

Bandwidth — How Much Information Can Be Carried?

Just as a large-diameter pipe can carry more water than a small one, so a signal spanning a wide band of frequencies can carry more information than can one using a narrow band. This range of frequencies is termed the "bandwidth." For example, a terrestrial TV message which is amplitude modulated and relayed in the frequency range from 66 to 72 MHz has a bandwidth of 6 MHz. This signal is spread onto a 36 MHz band in many types of satellite FM broadcasts.

Each type of communication medium requires a characteristic bandwidth. Television links need a substantially wider bandwidth than do audio messages such as radio or telephone because much more information is necessary to recreate a picture than to recreate music or a voice. To illustrate, television messages broadcasted by satellites typically require a video bandwidth of 4.2 MHz. In contrast, voice channels normally require a bandwidth of only 3,000 to 4,000 cycles per second (3 to 4 KHz) for quality sound reproduction.

Amplification and Attenuation

Man-made communications must often be amplified on the voyage from sender to receiver in order to keep the information intact because their power is usually weakened or attenuated. In the same fashion that a photograph is enlarged but not changed, correct amplification retains the original message. All receiving devices amplify a signal before demodulation occurs. For example, messages beamed into space from an uplink antenna are weakened on their voyage to a satellite as the signal becomes spread out and is absorbed by water vapor, clouds and other materials. The purpose of a satellite receiving antenna is to reamplify these weak signals, by collecting and concentrating them like a magnifying glass.

Occasionally, a signal will be intentionally attenuated. For example, a cable TV headend may deliver an excessively powerful signal to a feeder line which could overdrive connected TVs and cause distortion. Pads or line attenuators are then inserted to reduce the signal power.

Noise — Hindering Clear Communication

In a perfect communication system signals would be relayed with no interference or noise. However, television or radio broadcasts are occasionally of poor quality or "noisy." Noise is present in all matter at temperatures above absolute zero, 0 °K, the temperature at which all molecular motion ceases. There is no temperature colder than absolute zero! Note that °K is an abbreviation for degrees Kelvin above absolute zero. Kelvin degrees have the same magnitude as degrees Centigrade, they are just measured from a different starting point. Absolute zero, 0 °K equals minus 273.16 °C or minus 459.69 °F.

Noise is caused by the endless motion of the molecules that compose all matter. These small, vibrating charged particles generate electromagnetic waves that can mask the organized signal sent by man-made devices. Noise from the environment becomes stronger as the temperature increases. Therefore, satellite antennas inevitably detect noise from the warm ground. Furthermore, noise is generated by internal heat in amplifiers, receivers and other electronic equipment.

Random noise is always present in satellite communication systems. In fact, quality of a communication link is determined by the relative amounts of signal and noise power as measured by the signal-to-noise ratio. This is an indicator of how well the signal can be transmitted so that it is clearly stronger than the noise power. For example, if a signal of 1 volt is received along with 1 millivolt of noise, equal to an S/N of 1000, the picture quality would be poorer than if a signal of 2 volts is received with 4 millivolts of noise, an S/N of 500. Typically, televisions must detect a signal having power greater than 63,000 times the accompanying noise in order for a "high-quality" picture to be reconstructed.

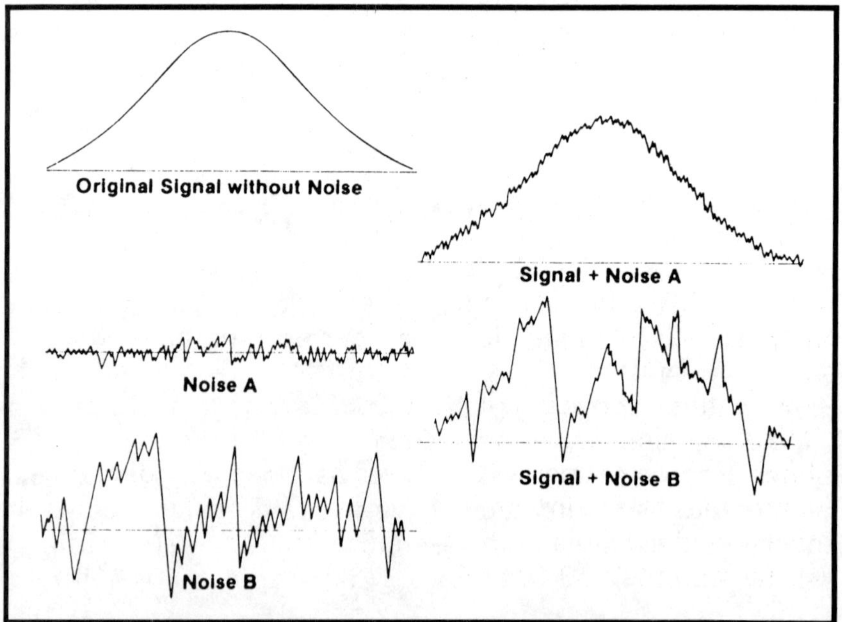

Figure 2-4. Noise. *The ratio of signal power to noise power, signal-to-noise ratio, defines the quality of a communication link. Excessive amounts of noise can completely garble an audio or video transmission so as to make it unintelligible.*

One principle advantage of digital relays is the signal-to-noise quality of the link. Repeaters along the communication route detect the presence or absence of a pulse, equivalent to either a one or a zero, and regenerate the pattern of pulses. Even a pulse which has been severely attenuated or distorted can usually be recognized as either a one or zero so that the original message can be perfectly restored. The audio component of scrambled satellite broadcasts is often digitally relayed and can therefore potentially be reproduced with higher fidelity than analog transmissions.

C. FREQUENCY ALLOCATIONS

The history of communication is unfolded by an exploration of how the frequency spectrum has been assigned. As technology progressed, man has become capable of recruiting ever higher frequencies. Wire transmissions were first relayed on relatively low frequency carriers, since the pioneers were limited in frequency by the rather primitive available electronics. Ironically, when radio waves above 1.5 megahertz frequency were first produced, they were delegated to the "hams" because, at that time regulatory bodies could not see a use for this region of the spectrum. Ham radio operators have contributed numerous innovations since that time. For example, the first home satellite receiving stations were built by hams. As technology progressed, coaxial cable transmission, microwave relays and then satellite communication were allocated successively higher frequencies.

The frequency spectrum was allocated up to only 30 MHz at the first World Administrative Radio Conference (WARC) in 1927. Since that time, these meetings which are sponsored by the International Telecommunications Union (ITU), an agency of the United Nations, have been held at regular intervals. Today, worldwide frequency allocations range up to 300 GHz.

It is important to understand that identical audio or video messages can be modulated onto and transmitted over any portion of the allocated spectrum. The frequency of the carrier wave is chosen given the type of communication channel as well as the available spectrum.

UNDERLYING CONCEPTS

In spite of the rapid development in man-made forms of communication, a relatively small portion of the electromagnetic frequency spectrum is presently in use. However, in those ranges where frequency space has been allocated, these resources are in heavy use and competition for assigned space can at times be fierce. As a result, innovative methods have been developed to "reuse" or to have more than one user simultaneously share the same portion of rare spectrum.

TABLE 2-2. NORTH AMERICAN ASSIGNMENT OF SOME RADIO FREQUENCIES

Frequency (MHz)	FCC Assignment
3-30	Wire services, CB, ham, etc (HF)
30-54	Mobile Radio
54-72	TV Channels 2-4 (VHF)
72-76	Radio Services
76-88	TV Channels 5 & 6 (VHF)
88-108	FM Radio
108-120	Aeronautical
120-136	Aeronautical
136-144	Government Use
144-148	Amateur Radio
148-151	Radio Navigation
151-174	Land, Mobile, Maritime
174-216	TV Channels 7-13 (VHF)
216-329	Government Use
329-420	Military
420-450	Ham
450-470	Mobile Radio
470-806	TV Channels 14-69 (UHF)

D. THE DECIBEL NOTATION

The decibel notation was created by engineers and scientists to describe extremely large changes in power, voltage and other parameters by small, manageable numbers. To illustrate, an amplification of 158,000 is expressed as +52.0 decibels. Or the severe attenuation of signals travelling the long path from a satellite to an earth receiving station can either be described by −196 decibels or by a factor of billions of billions.

Some examples of decibel (dB) changes are listed in Table 2-3. More details about the decibel notation are presented in Appendix C.

TABLE 2-3. THE DECIBEL NOTATION

Number of Decibels	Relative Increase in Power
0	1
1	1.26
3	2
10	10
20	100
30	1,000
50	100,000
100	10,000,000,000

III. PRINCIPLES OF TELEVISION

In order to comprehend all but the most basic scrambling method, it is imperative to clearly understand how a television processes a broadcast signal to recreate a picture and the accompanying sound. This study is a fascinating one which can be pursued in great depth by an electrical engineer following a dedicated career path. However, the purpose of this chapter is to explore the subject on a level at which a competent do-it-yourselfer is comfortable.

A. BACKGROUND

Television was considered a remarkable invention when it was formally introduced at the 1939 New York World's Fair. One of the first chronicled references to television had come in 1910 in the Kansas City Star. A story entitled "Television on the Way" described experiments by a French scientist. Some time then passed before C. Francis Jenkins, an American TV pioneer, predicted in 1922 that "motion pictures by radio in the home and an entire opera some day may be shown without the hindrance of muddy roads."

The early inventions that paved the way to a viable system were centered around the concept of "painting" an image. In 1884 Paul Nipkow, a German, invented the Nipkow disk, a device that dissected a scene into fragments of light. These were converted to electrical impulses in order to be relayed by radio waves. In order for these fragments to be reassembled into the matching image, it was necessary to develop a device to recreate the picture. The critical step leading to this breakthrough was the 1895 invention of the cathode-ray tube by Sir William Crookes, a British scientist who surprisingly considered his device merely a scientific plaything. The modern picture tube, developed by Vladimir Zworykin, is a sophisticated version of the Crookes tube. Although Crookes had used an electron beam to illuminate the surface of a phosphorescent-coated glass screen, Zworykin was the first to demonstrate control of this beam solely by electronic signals.

These pioneering developments led to the formal introduction of black and white television at the 1939 New York World's Fair. By the early 1960's color television was becoming commonplace.

B. THE TELEVISION PICTURE

The purpose of a television set is to recreate the picture and sound originally captured by a television camera and microphone as accurately as possible. A picture is "painted" line by line onto a television screen by an electron beam which regularly scans across its inside phosphor face. This is possible because the beam can be controlled by electrical fields at its source in the heart of the television tube. In order to properly recreate the picture, identical scanning must be employed at both the transmitter and receiver.

When no video signal is present, a pattern of scan lines called the "raster" is produced. The raster can be seen by tuning a television to an unused channel. However, when a signal is added to the raster, it increases and decreases the intensity of the electron beam. The more intense the electron beam, the brighter the illumination. The scanning caused by an organized television signal produces organized changes in illumination which are perceived as a picture.

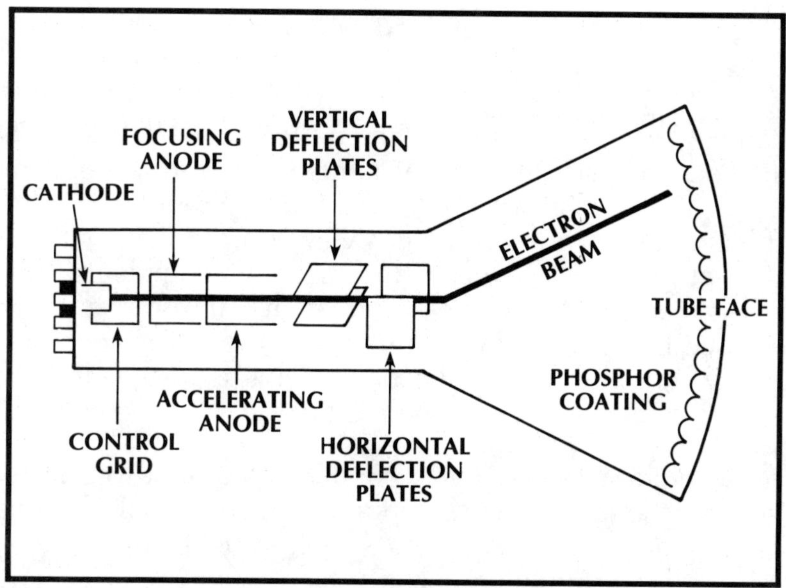

Figure 3-1. The Black and White Television Picture Tube. *A television picture tube produces illumination when a directed beam of electrons strikes the sensitive phosphor coating on its inner face. This beam is produced when electrons "boil off" a heated cathode, are directed by an electrically charged control grid, are focused by a positively charged anode and are accelerated by another positively charged metal plate. The scanning circuitry controls electrical fields on the vertical and horizontal deflection plates and therefore produces organized tracing. Screen illumination is controlled by the intensity of the electron beam.*

A black and white picture is composed by a single beam while a color picture is created by the scanning of three beams over three independent patterns of blue, green and red phosphor dots embedded into the screen surface. All other colors can be derived from these three basic ones.

Scanning

Scanning was the solution to the problem of recreating a complex scene that occurred simultaneously at many points in space, and transmitting it as a sequential stream of information. Scanning begins at the top left-hand corner of the screen as viewed from the front. The first line is swept across the screen. This "trace" is completed when the right side is reached. No picture information is transmitted during the retrace while the beam moves from right to left. The second line is then traced and so on. After the bottom line is traced, the beam is again turned off and is repositioned at the top of the screen.

The scanning process is very similar to that followed in reading a book where the reader's eyes start at the upper left hand corner and information is absorbed only during the left-to-right scans. The vertical retrace is similar to turning a page.

Aspect Ratio

Television pictures are transmitted in a rectangular form. The proportions of this rectangle are defined by a quantity known as the "aspect ratio." An aspect ratio of 4 to 3 was chosen. Therefore, if the picture is 4 inches wide it must be 3 inches high; if it is 12 inches wide it must be 9 inches high. Although these proportions can be adjusted to have a different aspect ratio via external knobs which change control voltages within the picture tube, the transmitted picture has minimum distortion at the standard 4 to 3 aspect ratio.

Vertical Resolution

The number of lines that are used to scan a picture determine its vertical resolution. Clearly, as the number of lines are increased, the system is capable of displaying a more detailed scene. As few as 405 and as many as 819 lines per frame had been used in the earlier days of television. The lower line number resulted in poor vertical resolution while the higher number required the use of an unacceptably large wide bandwidth. Today, either 525 or 625 scanning lines have become accepted in television systems around the world. For example, in North America 525 scan lines are used to recreate a television picture while in Europe 625 are standard. Note that some military cameras use 825 lines and high definition TV systems employ in excess of 1,100 scan lines.

Trace Rate and Horizontal Resolution

The choice of trace rate involved a tradeoff. Ideally, individual pictures or frames should be painted onto the screen as rapidly as possible in order to simulate continuous motion as closely as possible. However, at higher trace rates, the amount of brightness produced on the screen surface decreases because of inherent limitations in response rate of the phosphor coating. The beam would stay in one location for shorter periods of time as trace speed increases and would therefore have less effect. In addition, a higher frame rate also necessitates use of a wider transmission bandwidth because more variations in intensity are relayed.

The wider the channel bandwidth, the finer can be the changes in signal voltage and, therefore, the greater the number of brightness changes that may be transmitted in each line. The quality of the phosphor screens also determines how well a television can respond to the changes in signal bandwidth. Therefore, horizontal picture resolution is determined by transmission bandwidth as well as by the design and construction quality of the television set.

Frame Rate and Picture Stability

When television was first developed it was necessary to establish a repetition rate that would be high enough to simulate instantaneous information delivery. In other words, if a picture were painted onto the screen too slowly it would appear to flicker as in an old-time movie. It is no coincidence that movies were originally named "flickers." In the human eye a phenomena called persistence of vision occurs whereby visual images do not disappear immediately but diminish gradually. This is very noticeable, for example, when one looks at a flashbulb and sees an after-image. On the average, about 1/50th of a second is required for the perception of light to disappear entirely. Both the television and motion picture industry might not have been possible without this underlying human physiological characteristic.

Original black and white transmissions had their vertical timing, known as the frame frequency, set to the standard power line frequency of the country in question. Therefore, for example, North American televisions painted the picture 60 times each second while European sets had a frame frequency of 50 pictures each second. Synchronizing the frame and line frequencies had two advantages. A very stable frequency that was readily available in each country was used. This choice also eliminated the need to completely filter the line frequency in order to avoid the potential for an annoying type of interference known as hum bars. This interference shows up as a distorted rolling band that moves vertically through the picture. This would have been likely to occur if a different scanning frequency had been used and still can occur when the power is "dirty" and has ripples.

Interlacing

Engineering committees finally decided to use a frame rate of 30 pictures per second. Since this was a multiple of the 60 cycle line frequency, no problems with hum bar interference would be encountered. However, this relatively low frame rate would yield a picture with visible flicker. Choosing a frame frequency of 50 pictures per second would completely avoid any flicker but created a second problem. A signal bandwidth of at least 8 MHz would be required to relay 525 or 625 lines per frame at the minimum field frequency.

So an ingenious solution to provide the same resolution at half the bandwidth known as interlacing was invented. The first time the picture is scanned with only the even lines; the second scan paints the odd lines on the screen. Both fields have 262.5 lines for a total of

525 lines. This interlacing of even and odd effectively allows use of half the number of lines. This retains the visual illusion of an equivalent amount of flicker had all the lines been scanned at the same field frequency. Each of these half scans is called a field; two fields equals one complete picture or frame.

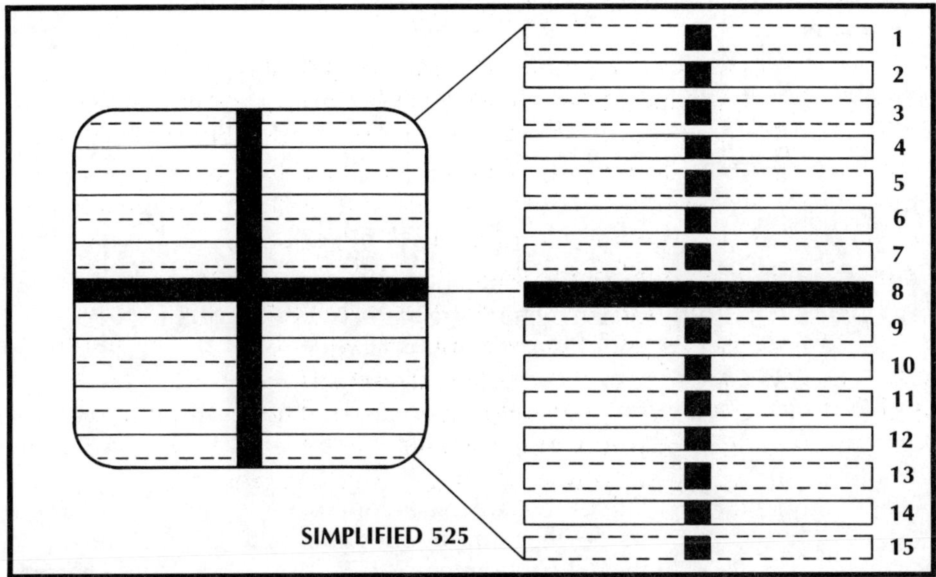

Figure 3-2. Scanning and Interlacing. *This diagram shows how a cross is "painted" onto the screen of a television tube. The scanning is interlaced because first the odds lines, 1,3,5,..., 15 are drawn before the even lines, 2,4,6...,14 are traced. This process happens so rapidly on a television screen that both fields appear to be simultaneously present.*

TABLE 3-1. THE BASIC PARAMETERS OF TELEVISION

Parameter	Effect	Value
Number of Scan Lines	Determines the vertical resolution. A finer picture results as the number of scan lines is increased.	525 lines 63.5 microseconds per line
Field Frequency	Determines the amount of flicker. The greater the field frequency, namely the greater the rate at which pictures are painted on a television screen, the lower the flicker and better picture stability. Brightness can decrease with too rapid a scanning rate.	30 per second 262.5 lines 16.7 milliseconds per field
Frame Frequency	Since there are two fields per frame, the frame frequency is half the field frequency.	60 per second 525 lines 33.4 milliseconds per frame
Channel Bandwidth	As channel bandwidth increases, a higher number of scan lines and a greater field or frame frequency can be accomodated.	6 MHz
Interlacing	Scanning half a frame or complete picture at once allows the flicker to remain constant but permits the bandwidth to be cut in half. Each half scan is a field.	

C. WORLDWIDE TELEVISION STANDARDS

In order to bring order into what had earlier been a more disordered world, three "standard" television broadcast systems have evolved. Although these standards all use a similar method of scanning, they differ in the number of lines in a frame and in the way that the color information is encoded. Their development was based on the power line frequency and the channel allocation scheme adopted in each country. The field frequency is matched to the line frequency which is typically either 50 or 60 Hertz.

TABLE 3-2. WORLD TELEVISION STANDARDS

ABBREVIATION	NAME	YEAR DEVELOPED
NTSC	National Television System Committee	1948
SECAM	Sequentiel à Memoire	1957
PAL	Phase Alternating Line	1961

The PAL and SECAM standards evolved from and were improvements upon the original NTSC format. The details of these standards and the various combinations used throughout the world are outlined in Appendices A and B and are referred to at pertinent places in the text. However, unless otherwise stated, all the examples used in this book will be drawn from the NTSC format used on the North American continent, among other places. Nevertheless, it is important to realize that all the concepts and conclusions about encryption systems discussed in this book are generally applicable to any television system used anywhere in the world.

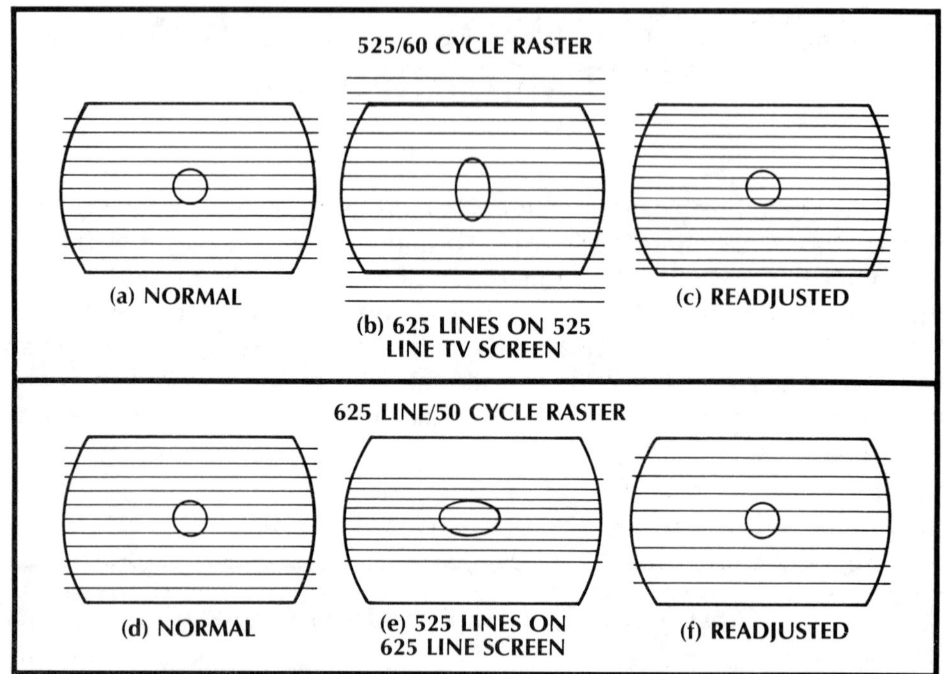

Figure 3-3. The Effect of Varying Scan Lines and Frame Frequency. *An normal NTSC raster of 525 lines scanned in two fields at 60 times per second is shown in (a). If a PAL broadcast having 625 lines at 50 fields per second is played on a NTSC set, even after adjusting the vertical hold so that the picture does not roll, the lines will not fit on the screen and distortion will occur as indicated in (b). In (c) the picture height has been reduced to restore the picture to normal. The same effect occurs in reverse as shown in (d), (e) and (f) when a NTSC transmission is played onto a PAL television receiver. In (f) the picture height has been expanded to accomodate the NTSC transmission.*

D. SCANNING THE BLACK AND WHITE TELEVISION PICTURE

Scanning of the television screen begins at its top left-hand corner. Following the horizontal scan of each line, the beam is turned off during retracing and repositioned at the beginning of the second line. This process is repeated until the whole field has been scanned. The period known as the horizontal blanking interval is necessary to prevent the beam from causing unwanted white lines during its retrace.

When the end of one picture is reached, after the end of line 262.5 which ends at the bottom center of the screen, the beam is again shut off during the vertical blanking interval so that it can be repositioned to the top of the screen for a restart of the scanning process without causing unwanted retrace lines. The second field scan begins at the top center of the screen. During the vertical blanking interval which lasts approximately 18 to 21 horizontal line periods, the shut-off beam is still following the horizontal trace. Of the 525 scanning lines per frame, only 483 are "active" and actually produce the picture. Other information such as captions for the hearing impaired, customer "addresses" for decoders or teletext can be inserted into the video signal during either the horizontal or vertical blanking intervals.

North American television uses 525 lines per frame which are scanned 30 times per second. Thus, 30 still pictures are painted onto the television screen each second. In order to eliminate flicker and improve upon the illusion of continuous motion, two fields each of which is half of the whole picture and is composed of half of 525 lines, or 262.5 lines, are alternately impressed or interlaced upon the screen 60 times each second. Two sequential fields thus equal one frame which paints a complete picture.

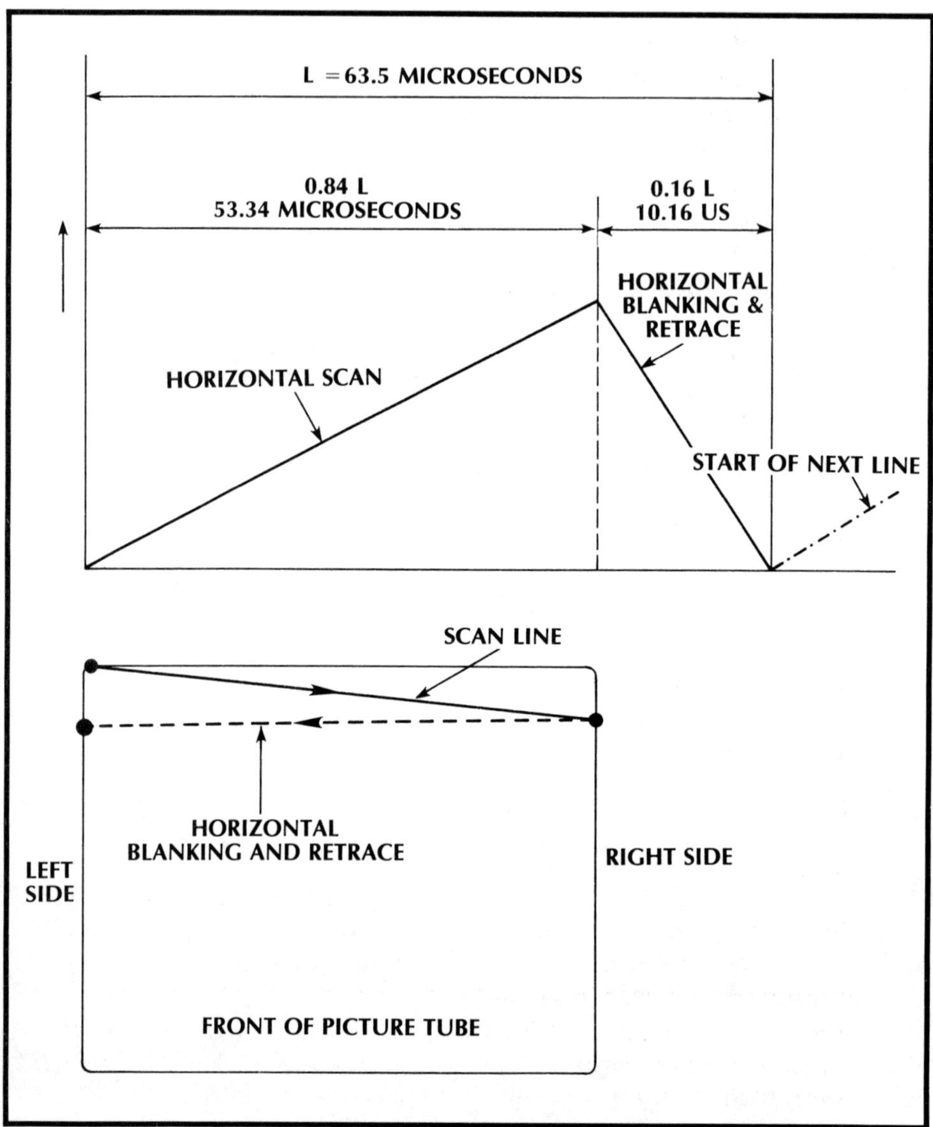

Figure 3-4. The Basics of Scanning. *The top diagram is a representation of the voltage applied to the horizontal deflection plates in a picture tube. This moves the beam from left to right across the screen while picture information is present and retraces the beam during the horizontal blanking interval when no illumination is produced.*

The total number of lines traced each second, which is known as the horizontal scanning rate, is therefore 15,750. This equals 60 times 262.5 or 30 times 525. Each complete line lasts 1/15,750th of a second, 63.5 microseconds. This 63.5 millionth of a second timing interval, L, is used to characterize the traces of the video signal shown in the accompanying figures.

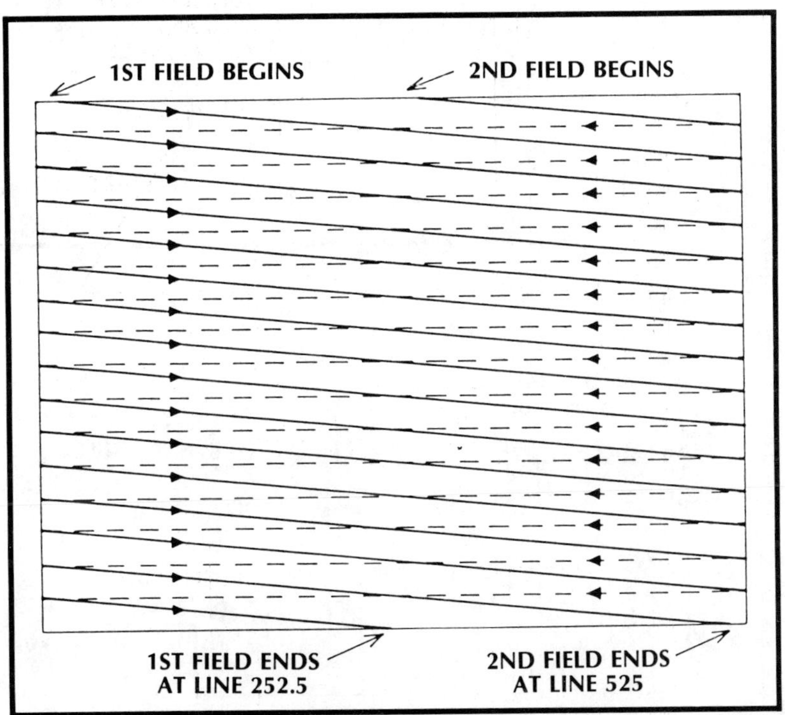

Figure 3-5. Scanning NTSC Fields. *The scanning of the first field begins at the upper left hand corner of the television tube and ends at the bottom center at line 252.5. The second field begins at at the top center and ends at the bottom right after line 525 is completed. Two of these fields equal one frame or complete picture.*

E. THE BLACK AND WHITE TELEVISION SIGNAL

The amplitude of the intelligence portion of the composite baseband video carrier varies with the illumination pattern in each scanned line produced by a television camera. Two reference levels are built into this raw video signal, the reference white and the reference black level. At voltages below the reference white level, the electron beam intensity is at a maximum level and the phosphor screen is as bright as possible. At levels above the reference black level, no illumination is produced.

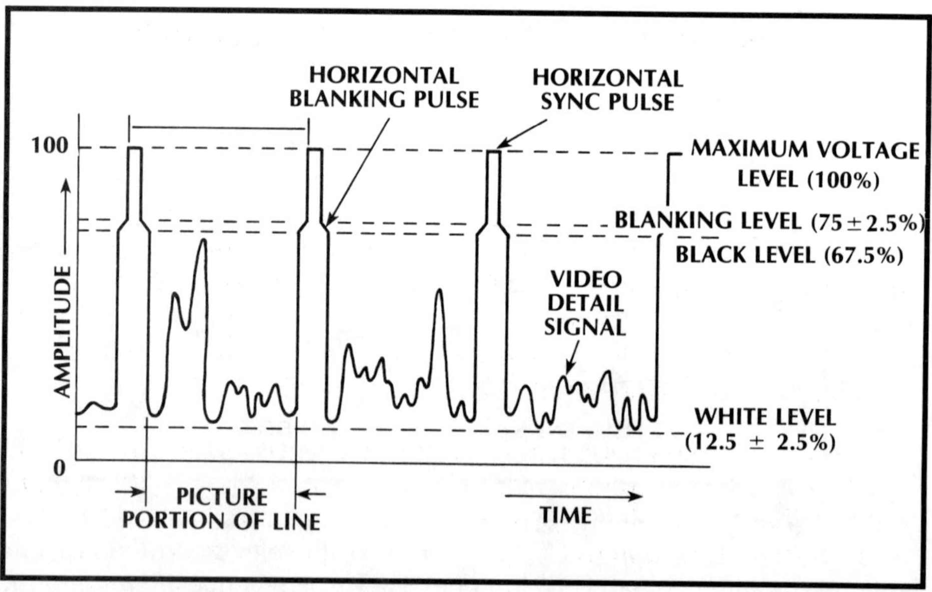

Figure 3-6. The Television Signal. *This illustration shows the structure of three lines of a television video signal. Above the 67.5% level, the black level, the beam is shut off. The "blacker than black" level falls between 67.5% and 100% of the maximum voltage level. The picture information is relayed between the black and white levels. Therefore, during the horizontal blanking interval, which contains the blanking pulse and sync pulse, no illumination is produced.*

This format whereby the black reference level is at a positive voltage on the modulated carrier is known as negative scanning. Other television systems use positive scanning where the signal has an inverted voltage pattern. Negative scanning has two important advantages. Since lower voltage signals are closer to the noise floor and more susceptible to noise, if the bright scenes which mask the noise are relayed at low voltage, noise is less visible. Also, in negative scanning, the sync is at the maximum voltage point so that if the signal happened to be somewhat compressed the picture information would be less affected than if the white level had been at the top of the waveform. Note that the transmission power is lower during the white portions of the picture when using negative scanning since voltages are lower. The opposite is true for positive scanning.

Synchronization and Blanking Pulses

In order to accurately recreate a television picture, pulses to synchronize the precise timing of the scanning are inserted onto the non-picture portions of the video signal at the television camera. The horizontal sync pulse sets the start of the horizontal trace. The vertical sync pulses are also inserted onto the signal between fields to correctly align the timing for the beginning of the vertical scan. If the vertical sync is not accurate, the television picture will "roll." If the horizontal sync pulse is removed or faulty, the picture will "tear."

The horizontal sync pulse occurs within the time period of the horizontal blanking interval. The blanking pulse begins at the end of the first line to shut the beam off. Then the sync pulse tells the beam to begin the second line. During this repositioning period the beam is still blanked off. Finally the second scan line begins. This continues until the end of the last scan line when the vertical blanking pulse begins. The vertical blanking interval lasts longer than the horizontal (18 to 21 horizontal line periods) because a longer time is required to move the beam during the vertical retrace from the bottom to the top. All non-picture portions of the video signal are at voltage levels above the black reference level and therefore do not have any visible effects.

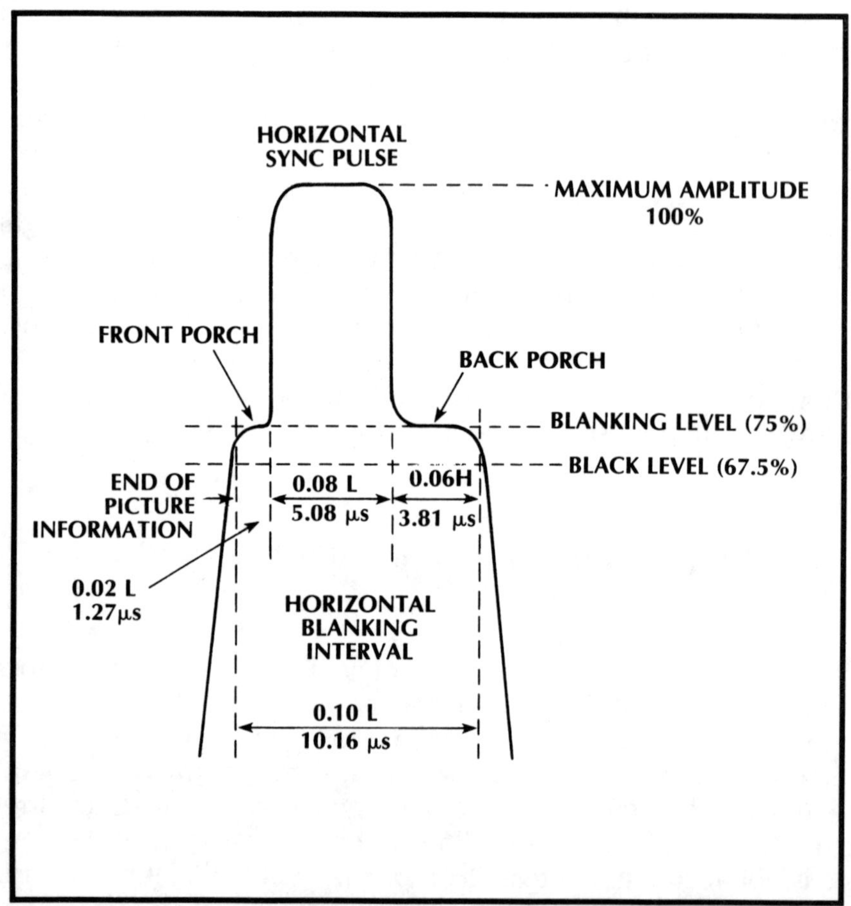

Figure 3-7. Details of the Horizontal Sync Pulse and Blanking Interval. *The structure of the horizontal blanking interval is diagramed here. The front porch blanks out the electron beam and signals the end of the picture information. The beginning of the horizontal sync pulse commands the circuitry to start the retrace. During the back porch time period, the beam is blacked out just before the start of the next line. If the television picture were to be moved either left or right on the screen both the front and back porch would be seen as vertical black bars running down both sides of the picture. Note that the interval L is equal to the line duration of 63.56 microseconds (which equals one divided by the line scanning frequency, 15,734 Hertz for color transmissions).*

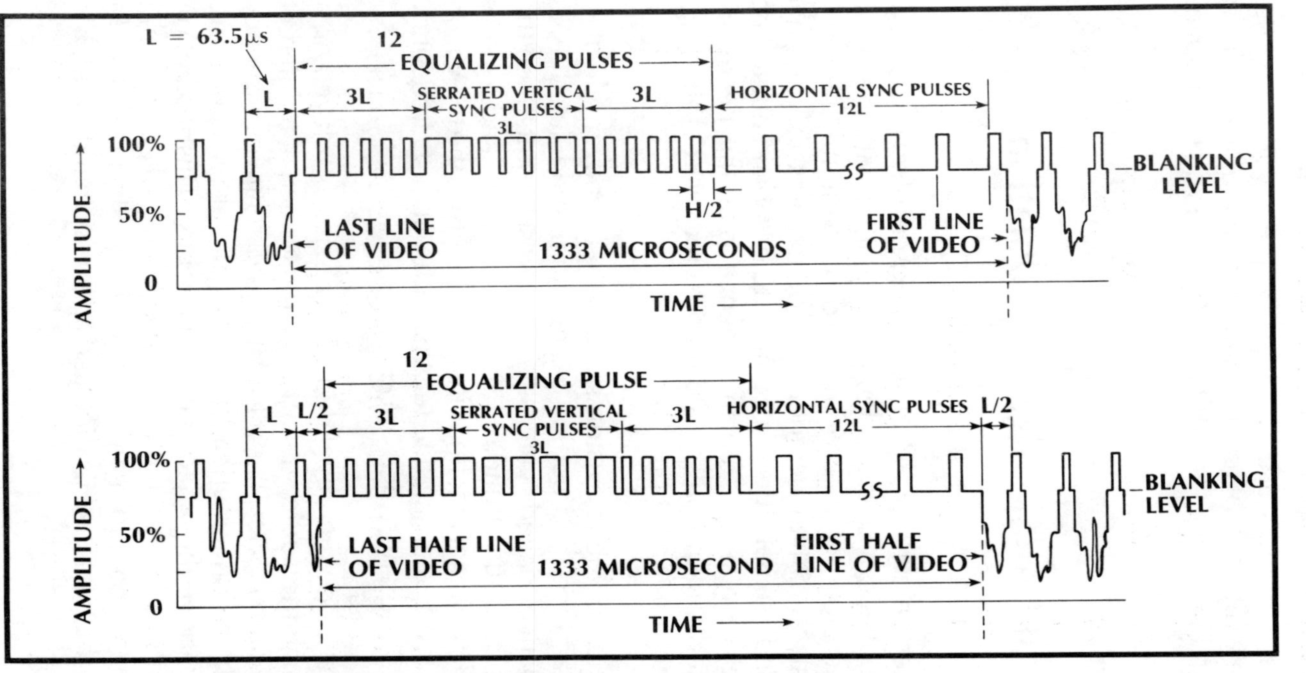

Figure 3-8. Details of the Vertical Sync Pulse and Blanking Interval. *The top diagram shows the vertical interval following the last trace line of one frame and the first of the subsequent one. The bottom figure show how the first field ends at the bottom middle of the screen and the second starts at the top middle. The total time for the entire vertical interval is 1333 microseconds. The vertical retrace begins with the first vertical sync pulse. Serrated vertical pulses are necessary so that both the vertical and horizontal sync is maintained during the vertical blanking interval. These and the equalization pulses are required so that television receiver electronics functions properly. In newer digital televisions equalizing pulse are redundant.*

More details about the video signal are presented in the accompanying figures. The 100 units used to describe the 1 volt swing between the reference white and black level are referred to as IREs (after the Institute of Radio Engineers). The full range of the sync pulse spans 40 IREs.

Video and Audio Signal Modulation

The broadcasted television signal is composed of two separate carriers, one is modulated with the video and the other with the audio information. A television set receives both carriers simultaneously, amplifies the signals and then demodulates or extracts the original "composite baseband" signal. It is the composite baseband audio and video signals which contain all the intelligence plus the necessary synchronization and blanking pulses required to reconstruct the original picture and sound information.

The audio portion of the television signal is frequency modulated (FM) to reduce noise; the video portion is amplitude modulated (AM) to minimize disturbances called "ghosts." Such ghosts result when the video signal is reflected off a building or another object and follows a different path from the signal that finds its way directly to the television antenna. Note, however, that satellite television broadcasts frequency modulate both the audio and video information.

The audio associated with any video transmission is relayed on a "subcarrier." Although the audio subcarrier is transmitted at a different frequency than the video carrier, it is offset at a fixed relative frequency on each channel. In North America the audio subcarrier center frequency is 4.5 MHz above the video center frequency. This separation was chosen so that the audio would be far enough removed from and not interfere with the video information but so that both signals would be contained within the total 6 MHz allocated television bandwidth.

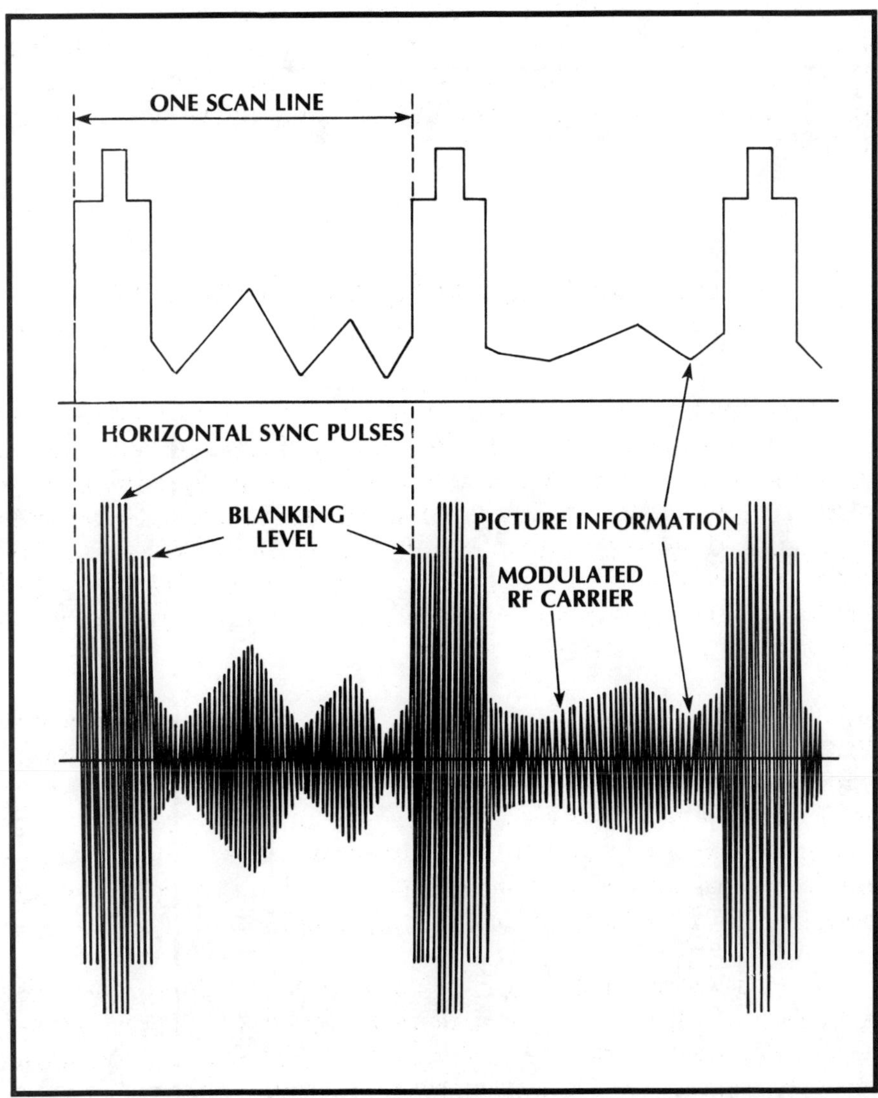

Figure 3-9. Video Signal Modulation. *The video signal is amplitude modulated onto the carrier wave. In this case, the blacker levels are modulated less than the whiter levels and therefore cause less changes in the voltage of the carrier. This is known as negative scanning.*

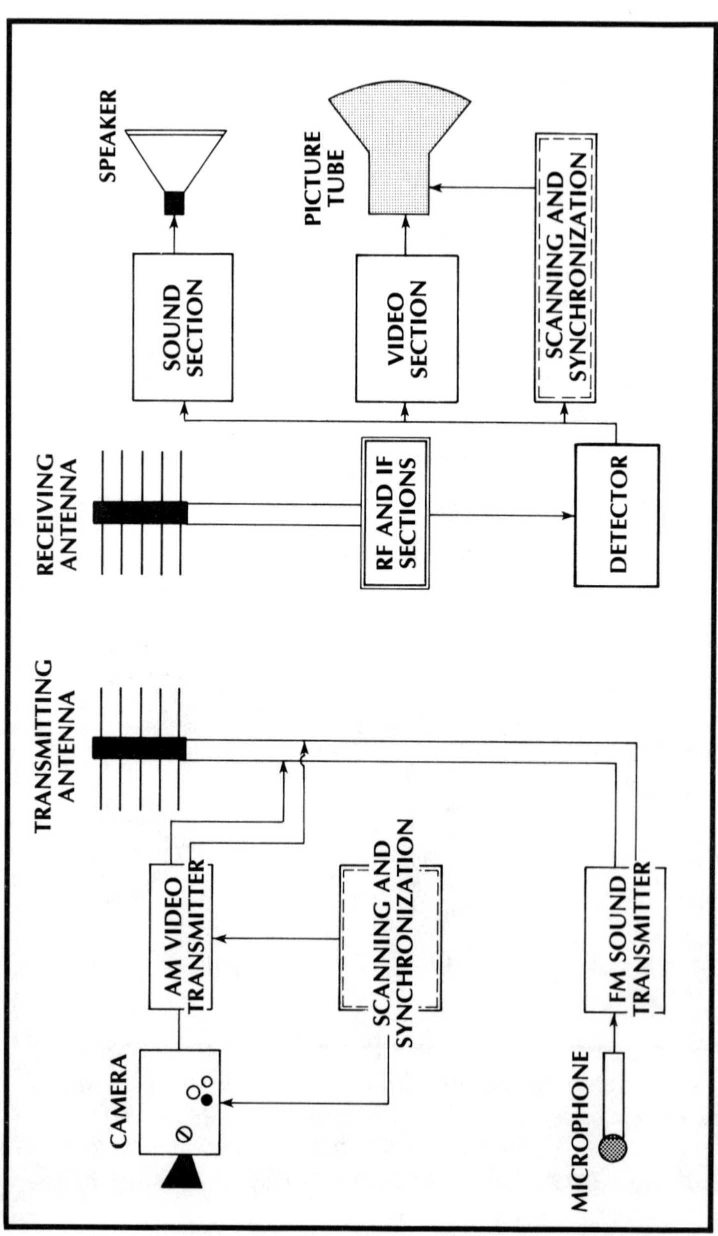

Figure 3-10. The Basic Television Circuit. *The same scanning and synchronization pulses that drive the television camera are also added onto the video signal for modulation. AM is used for over-the-air and cable broadcasts while FM is used for satellite transmissions. This signal is transmitted with the FM audio to a receiving antenna. The television receiver demodulates both signals and feeds them to audio and video sections. Scanning and synchronization information is extracted and is used to drive the electron beam. The beam intensity is controlled by the video section.*

Channel Bandwidth Assignments

North American VHF (very high frequency) and UHF (ultra-high frequency) channels are each assigned a 6 MHz wide bandwidth. There are a total of 82 plus allocated channels, 12 of which are in the VHF band with the remainder being in the UHF band (see Appendix B for more details). There are many other associated cable TV channels in the midband range (A2, A1 and A through I, 108 to 174 MHz), superband (J through W, 216 to 300 MHz) and hyperband (AA through ZZ, 300 to 456). Note that these cable TV bands are accessible only via cable converters or cable ready TV's or VCR's.

TABLE 3-3. NORTH AMERICAN CHANNEL ASSIGNMENTS

	OFF-AIR CHANNELS	
Band Designation	**Channel Numbers**	**Frequency Range (MHz)**
Lo-VHF	2 - 6	54 - 88
FM Band		88 - 108
Hi-VHF	7 - 13	174 - 216
UHF	14 - 69	470 - 806
	CABLE TV CHANNELS	
Sub-VHF	T7 - T13	5.75 - 47.75
Lo-VHF	2 - 6	54 - 88
Mid-Band	98(A2) - 99(A1)	108 - 120
	14(A) - 22(I)	120 - 174
Hi-VHF	7 - 13	174 - 216
Superband	23(J) - 36(W)	216 - 222
Hyperband	37(AA) - 62(ZZ)	300 - 456
	63 - 94	456 - 648

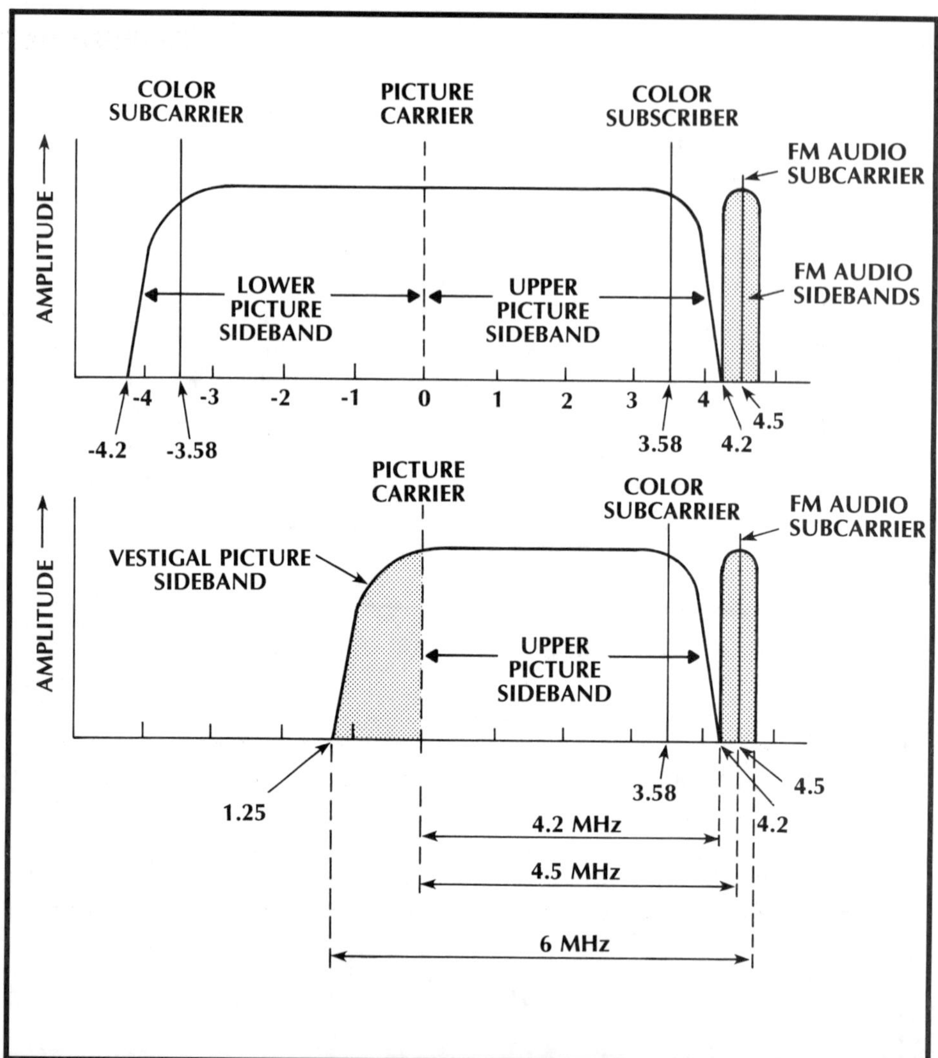

Figure 3-11. Structure of an NTSC Television Channel. *The process of amplitude modulating a video signal results in creation of an upper and lower frequency sideband. For example, if a 5 KHz signal is frequency modulated onto a 100 KHz carrier, two distinct sidebands at 95 and 105 KHz are produced. Both sidebands in any signal contain all the necessary intelligence to recreate the original information. Television transmitters suppress the lower sideband in order to reduce the required bandwidth. This is termed vestigal sideband modulation. But the vestigal sideband is required for complete demodulation. In the NTSC signal, the picture and audio carrier center frequencies are separated by 4.5 MHz, the picture and color subcarriers by 3.58 MHz. Both FM audio sidebands are transmitted within a total 50 KHz bandwidth.*

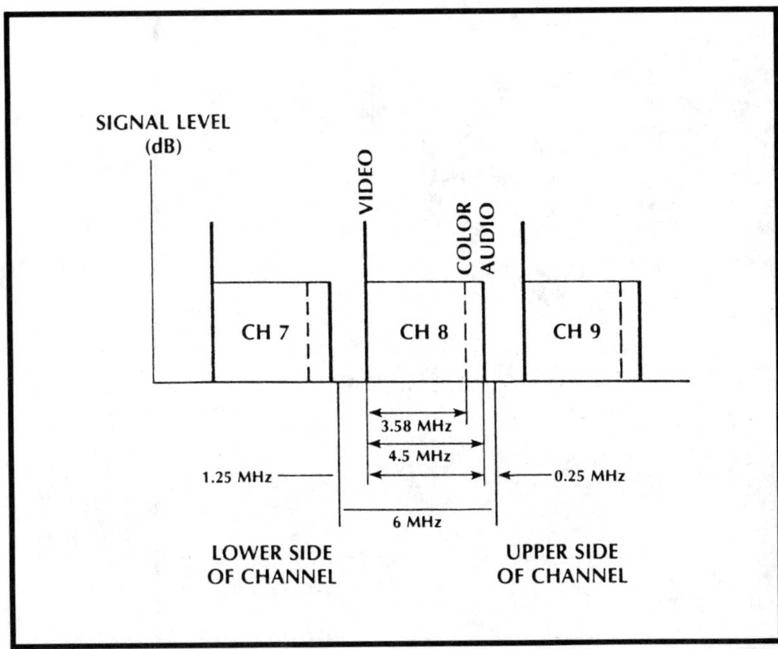

Figure 3-12. Adjacent NTSC Channels. *Each television channel is organized into a picture carrier centered 1.25 MHz above the lower side of the channel, a color subcarrier centered 3.58 MHz above the video and an audio subcarrier 0.25 MHz centered below the upper edge of the channel. The level of all three carriers are adjusted to minimize interference between each other and between those in adjacent channels.*

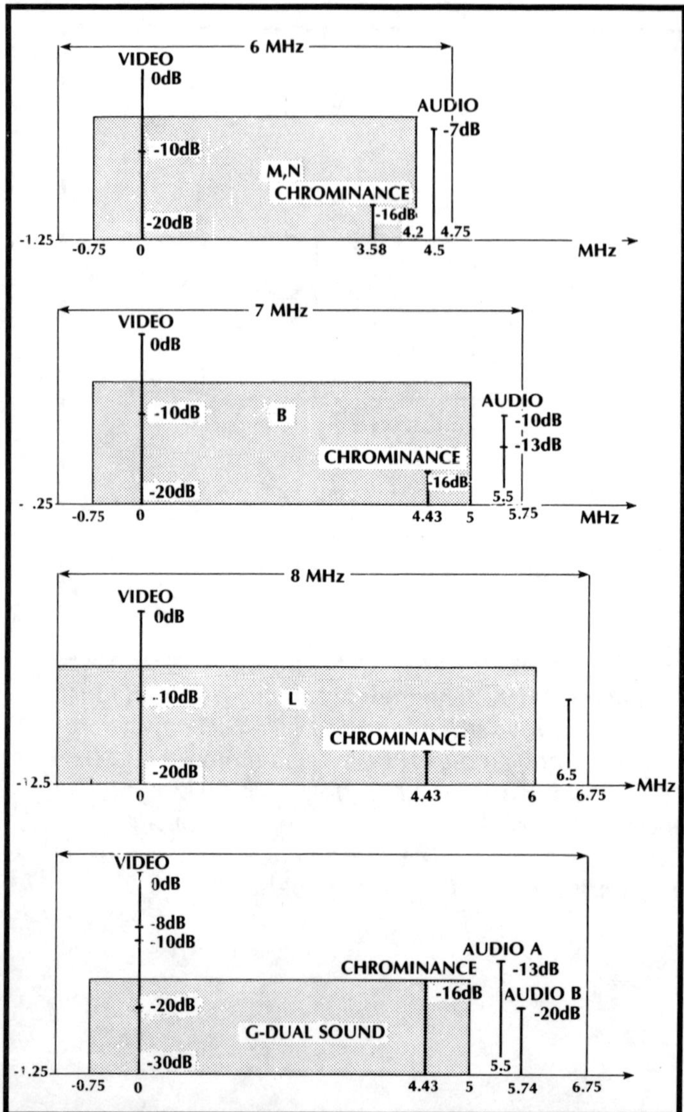

Figure 3-13. Comparison of Worldwide Channel Structures. The basic 6, 7 and 8 MHz wide channels used in NTSC, PAL and SECAM trasmissions are compared here. The levels of the various components are adjusted to minimize interference between each other and between adjacent channels. Appendices A and B outline the codes in use here.

Within a standard channel, the FM audio intelligence carrier has a center frequency 0.25 MHz below the upper edge of the channel and a bandwidth of approximately 50 KHz. This leaves 5.7 MHz available for the video carrier. The center frequency of the video carrier is then placed 1.25 MHz above the lower edge of the channel. Therefore, the audio carrier is located at 4.5 MHz above the video carrier. Since most of the video information is contained at higher frequencies above the video carrier, in effect, the composite baseband audio and video carriers occupy a band of frequencies ranging from 1.25 MHz to 5.75 MHz plus 50 KHz or just under 4.6 MHz. The reason for concentrating the signal in the middle of the 6 MHz bandwidth is to minimize adjacent channel interference.

G. THE COLOR TELEVISION SIGNAL

The complete color television signal is also composed of an FM sound carrier and an AM video carrier, both contained in a 6 MHz bandwidth. The video portion has the same sequencing of blanking interval and synchronization pulses as well as number of scan lines, field frequency and horizontal scanning rate. This similarity is essential since the color signal must recreate a black and white picture on a b/w television set. Both systems were designed to be compatible with each other.

The complete color signal is necessarily more complex than the b/w signal. The amplitude variations of the b/w video signal represent variations from complete darkness to bright white images. But the amplitude variations of the color signal must represent both the illumination and colors of the image originally captured by the television camera. In addition, the color signal includes a "color sync burst" which is 8 to 11 cycles of a 3.58 MHz sine wave. The color burst is inserted immediately after the horizontal sync pulse on the horizontal blanking pulse and serves to ensure that the recreated colors on the TV set match the initial scene at the studio. The color does not interfere with the horizontal sync control because it follows this horizontal pulse at half its peak voltage.

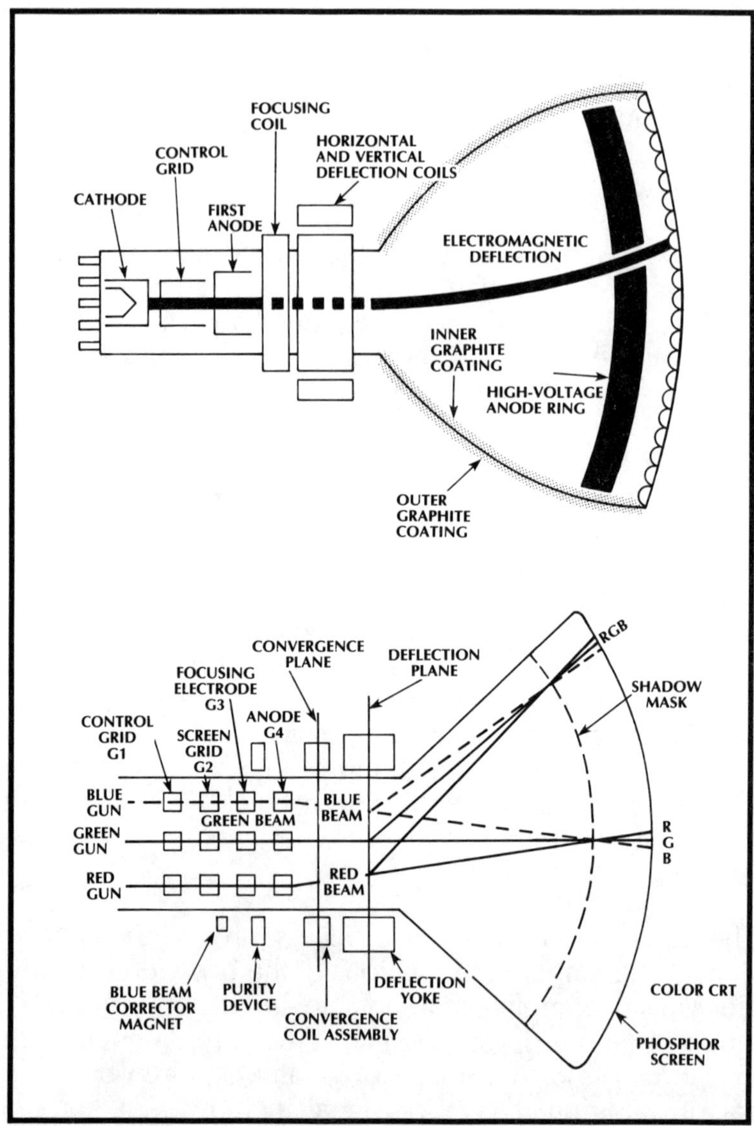

Figure 3-14. The Color Television Picture Tube. *A color television tube operates in much the fashion as a black/white tube except that three beams are used to scan the green, red and blue phosphor masks laid onto the screen.*

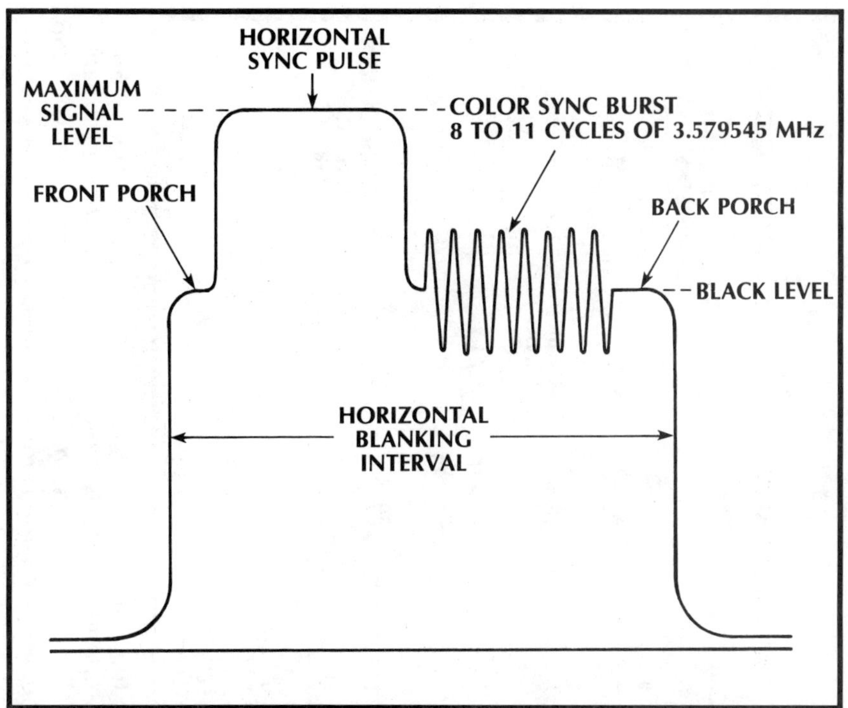

Figure 3-15. The Color Burst. *The color burst, 8 to 11 cycles of a 3.579545 MHz reference signal, is inserted onto the back porch of the horizontal blanking interval. It follows the horizontal sync pulse at half its voltage level and therefore does not effect synchronization. The color burst is a reference waveform which is essential in separating the raw color information from the difference signals.*

A color camera breaks a scene down into a signal for each of the three primary colors, red, green and blue, from which all other colors can be reconstructed. The luminance (Y) signal, which corresponds to the intensity of the original picture and from which a b/w scene can be recreated, is also assembled from all three color signals. The color information is translated into the I and Q signals, each of which is composed of two of the three primary color signals. These I and Q signals, also known as the color difference signals, are both modulated onto two 3.58 MHz subcarriers at 90 degrees out of phase with each other and then combined into the chrominance (C) signal.

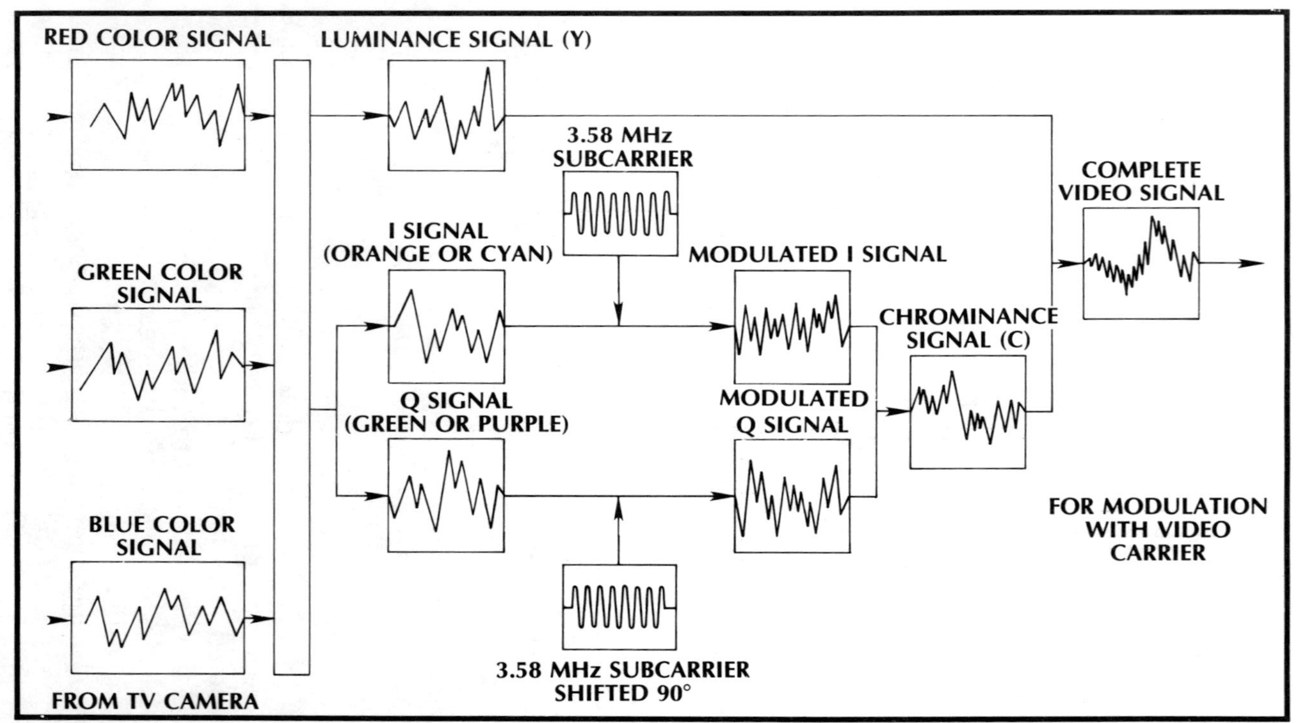

Figure 3-16. The Color Television Signal. *The three raw color signals from a television camera are combined to create the luminance signal (L) and difference signals (I and Q). The luminance information is sufficient to recreate a black/white picture. The difference signals are modulated onto two 3.58 MHz carriers having phases differing by 90 degrees and then are combined to form the color or chrominance signal. The chrominance and luminance signals are mixed together into the baseband composite video signal for modulation onto a transmission carrier. The 3.58 MHz subcarrier must be relayed along with the composite signal so that the raw color signals can be properly extracted from the difference signals.*

The luminance and chrominance signals are added together or frequency multiplexed to form the composite bandband video signal. This is amplitude modulated onto the carrier for transmission to a television set.

The television receiver uses the 3.58 MHz color burst, which is purposely chosen to have frequency identical to that of the color subcarrier, as a reference signal to reconstruct the I and Q waveforms from the chrominance signal. The phase of the chrominance signal relative to the color burst is critical to reassembling the color difference signals and in creating a picture true to the original. The television displays incorrect shades of color when the phase shifts from its correct value. Color distortion is also caused by incorrect changes in the amplitude of the subcarrier.

Color Scan and Field Frequencies

The field frequency chosen for color television scanning is 59.95 fields per second which is lower than the 60 Hz rate used in black and white television. This rate times the number of lines per field equals a scan frequency of 15,734.26 lines per second. The reason for this choice is related to the position of the color subcarrier relative to the center frequency of the luminance signal. The relative difference between these two frequencies could cause objectional interference unless it is related to a multiple of the horizontal scan frequency. The color television engineering committee slightly altered the scan and field frequencies to minimize this interference problem.

The difference between the 60 hertz line frequency and the 59.95 hertz field frequency is 0.06 cycles per second. This is low enough so that any hum bars produced that travel across the screen at one every 17 seconds (1/0.06) are not perceptible.

The Color Subcarrier and Frequency Interlacing

The center frequency of the color subcarrier is carefully chosen in order not to interfere with the luminance information. This choice is related to the line scanning frequency. Most of the black and white information is centered on multiples of the most prevalent frequency, the line scanning frequency. The color subcarrier frequency is then chosen to position the color information in the empty gaps between the concentrations of b/w energy. This is called frequency interlacing and results in a very efficient use of the available frequency spectrum.

The number of lines and the field scanning frequency differ between the PAL and NTSC transmission schemes. The video bandwidth in a PAL signal at 7 MHz is also wider than the 6 MHz one used in a NTSC signal. Therefore, in order to achieve the most efficient frequency interlacing, each standard has the color subcarrier at a different center frequency. While NTSC subcarriers are centered on 3.58 MHz, the PAL is at 4.43 MHz. When receiving PAL on an NTSC television, no color is seen and the picture rolls.

Color Distortion

Television receivers can be characterized by the fidelity of reproduction. Distortion of color hues, brightness, fine picture detail as well as other types of irregularities can occur at various stages of transmission, reception and processing.

Color television is especially vulnerable to differential phase and differential gain. Differential phase is a measure of the phase distortion that results when the luminance signal is varied from its black to white reference levels. It is expressed by the number of degrees of phase error experienced over this range. Since the colors that are recreated on a television screen are determined by the precise phase relationship between the color burst and the color subcarrier, differential phase errors can cause annoying changes in hue. (An instrument called a vector scope can be used to measure phase relationships.).

Differential gain is measured by the change in color subcarrier amplitude as the luminance is varied from its minimum to maximum value. It is expressed by the percentage that this unwanted change is relative to total signal amplitude. Excessive amounts of differential gain can cause some colors to look washed out, while others might appear to be excessively bright. These distortions would be caused by changes in the luminance signal resulting from brightness differences between scenes.

Fully saturated colors having a high amplitude chrominance signal require a greater bandwidth for high fidelity transmission than any other component of the television signal. Satellite receivers often are designed so that transmission bandwidth is reduced in order to lower noise and to improve receiver threshold, namely its ability to detect low signal-to-noise ratios. In the process, this can cause chroma "sparklies" to appear. Sparklies are more technically known as impulse noise. Lower cost and marginally designed satellite receiver brands can experience serious problems in reproducing saturated colors.

This need for a wide bandwidth is a reflection of the need for a good high frequency response. A system with good high frequency response can follow rapid changes in signal voltages and can produce crisp, clear pictures. As described in Chapter IV, digital audio signals and addressing information are transmitted by some encryption systems in the blanking intervals. If the frequency response is poor, ones can appear as zeros and vice versa. The result is "crackles" in the audio and the possibility of misreading subscriber addresses.

PAL and SECAM Solutions to Color Distortion

The PAL system was an offshoot of the original NTSC standard that was slightly altered to minimize the effects of phase distortion on color fidelity. The phase of the color signal is inverted from line to line. As a result, any phase errors that occur will have an opposite effect on sequential lines and the viewer's eye will average out and eliminate moderate amounts of color distortion.

The SECAM system transmits only one color difference signal per line. Two lines must be used to recover the color information so that some averaging also occurs. The MAC system outlined in the next chapter also uses sequential line transmission of the two color difference signals.

H. THE BASIC TELEVISION RECEIVER

The basic television receiver can be understood by examining its functional blocks and by following the signal flow.

A signal enters via either the VHF or UHF terminals and is connected to the appropriate tuner. This circuit component, whether controlled by a rotary switch, push button selector or an electronic remote, is responsible for selecting one channel or frequency band. A VHF tuner must be capable of spanning the range from 54 to 216 MHz or from 54 to 400 MHz for cable-ready sets. A UHF tuner accepts frequencies from 470 to 806 MHz.

The output of either the VHF or UHF tuner is a common intermediate frequency (IF) of 45.75 MHz centered on any selected 6 MHz channel. This IF is bandpass filtered to remove any unwanted signals from adjacent channels.

The signal is fed to a detector or demodulator which extracts the composite baseband signal. The detector also produces an automatic gain control (AGC) feedback signal that maintains the gain of the tuner section at the appropriate level so the detector is fed the proper voltage. The signal is then filtered to remove the 4.5 MHz audio subcarrier.

Prior to filtering, a portion of this video signal is tapped off and sent to the audio detection circuitry. Once the 4.5 MHz subcarrier is isolated and demodulated, it is filtered and amplified. Finally, a variable attenuator is used for controlling the voltage which drives a speaker.

The "clean" composite video baseband signal then enters a chroma or color demodulator which separates the difference signals into the three basic color signal components. These red, green and blue signals drive the television tube electron beams and display the picture.

A sample of this video signal is also fed to the synchronizing circuits. The timing pulses, which are an integral part of the signal, are stripped away in the sync separator. These send the correct voltages to the deflection circuitry to produce the organized picture scanning.

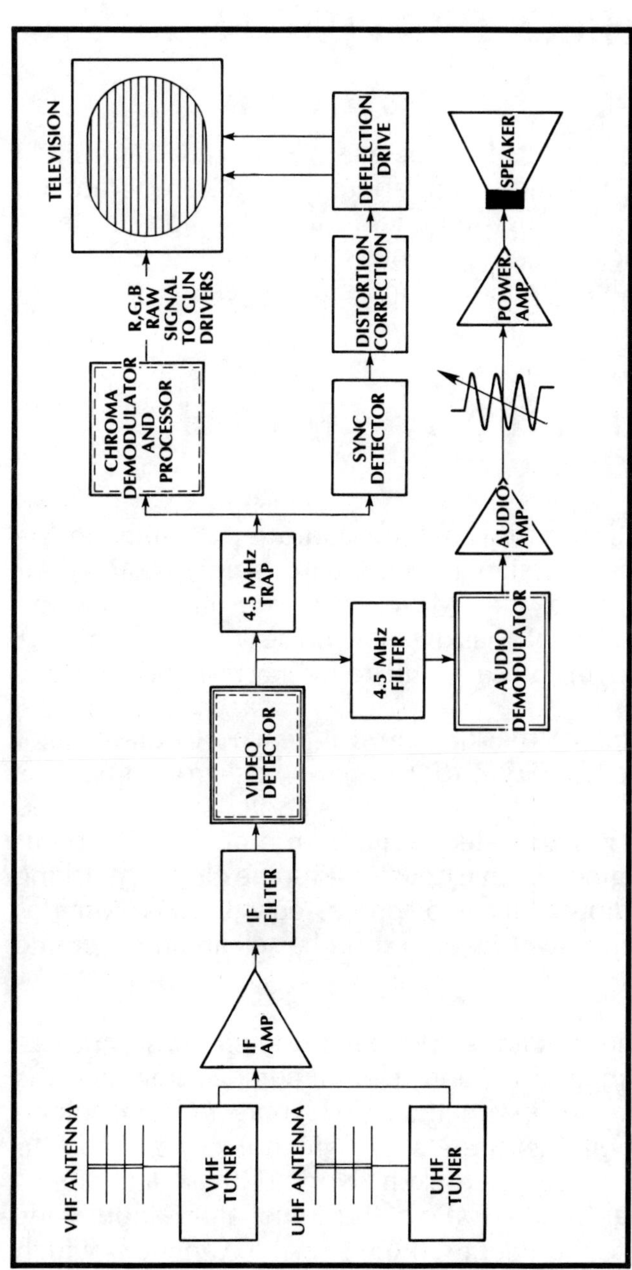

Figure 3-17. The Basic Television Receiver. *The operation of the basic television receiver is outlined in more detail in the accompanying text.*

I. STEREO TELEVISION

The rapid growth in satellite TV broadcasts which are transmitted with high fidelity stereo and the recent approval of a stereo standard for over-the-air television has focused interest and attention on stereo television. The techniques and equipment used in satellite, cable and over-the-air television have similarities and differences.

Transmitting High Fidelity Sound

One primary objective in transmitting audio is to minimize the amount of noise added to the signal en route to its final destination. Due to physical limitations of the FM link, this is usually supported by a second objective, that of masking the noise which cannot be avoided. Most of the annoying noise is high frequency hiss and static.

The type and amount of masking used is related to the type of audio being transmitted. Therefore, for example, background noise is much more noticeable if the program content is a soft bass note played by a cello rather than a raucous electric guitar run on a rock album. Over the years sound engineers have developed some elegant masking techniques so that an audio signal can be processed within the confines of a given bandwidth and power level to sound as clean and accurate as possible.

Three related techniques can be identified: pre-emphasis and de-emphasis; amplitude companding; and spectral companding which is a more general form of the first method. There are presently three patented companding methods: the Dolby, which is quite familar to audiophiles; the dbx; and the Telefunken. None of these techniques improve the overall signal-to-noise ratio of the audio information. They very effectively reallocate the energy in the signal to a pattern which matches that perceived as clean, high fidelity sound by the human ear.

Pre-Emphasis and De-Emphasis

Pre-emphasis was developed in conjunction with FM radio to counteract the increasing influence of noise at higher frequencies. A signal-to-noise ratio of 60 dB at low audio frequencies may be degraded by more than 10 dB in the higher frequency portions of the audible spectrum. In order to remedy the situation, higher frequency components of the audio signal are boosted in level before transmission. Then the noise detected during transmission is lower relative to the signal. De-emphasis, the mirror opposite of pre-emphasis processing whereby higher frequency components are reduced in level by the same amount, restores the original signal.

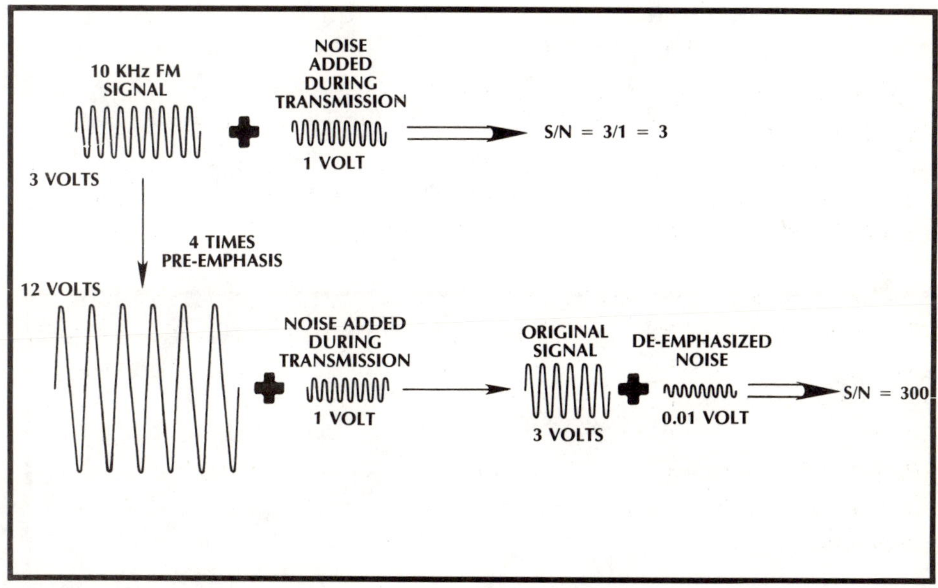

Figure 3-18. Pre-Emphasis and De-Emphasis. *This illustrates how the signal-to-noise ratio can be markedly improved by pre-emphasizing before transmission and de-emphasizing after transmission. The amount of emphasis applied to the higher frequency components of an FM signal is progressively increased to counteract the tendency towards reductions in signal-to-noise ratios.*

All FM radio broadcasts use the standard "75 microsecond" pre-emphasis. This time factor simply defines the relative amount of level boosting that occurs over frequencies within the audio spectrum.

Amplitude Companding

Pre-emphasis is activated only when high frequency information is present and has no effect on masking noise for low frequency, low amplitude signals which are not far above the noise floor. A method to improve perceived audio fidelity is to compress signals at the transmitting end. This means that the higher voltage portions of the signal are attenuated while those nearer zero amplitude are increased in level. This has the effect of beefing up the weaker parts of the signal so that the dynamic range is compressed or "squeezed." A compressed signal has a higher average level, and therefore may have more apparent loudness than an uncompressed one, even though the peaks are no higher in value. To illustrate, the patented dbx method typically reduces signals at maximum amplitude by 30%, leaves signals at 8.99% of peak unchanged, and boosts signals of 1% and 0.1% of peak by 3% and 95%, respectively.

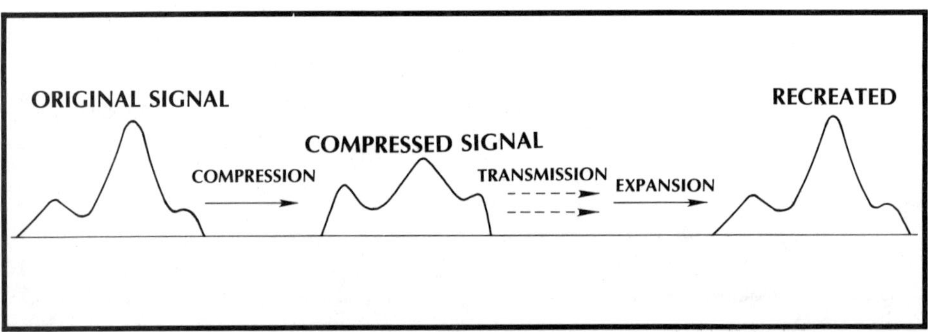

Figure 3-19. Amplitude Companding. *A signal can be compressed so that lower voltage portions are increased and higher voltage portions are decreased in order to increase the signal-to-noise ratio of those portions nearer the noise floor. This is a commonly used noise reduction technique in audio broadcasts.*

The mirror amount of expansion applied at the receiver restores the original signal. This chain of compression and expansion is known as companding.

Spectral Companding

When using fixed pre-emphasis alone, two potential difficulties still remain. First, a fixed amount of pre-emphasis applied to those audio signals containing predominantly high frequencies could boost them too much. This is known to audio engineers as lack of headroom, namely insufficient capacity for the peaks of the program to be cleanly tranmsitted. Second, the fixed pre-emphasis could be insufficient to adequately boost the very low amplitude, high frequency signals and therefore may not sufficiently mask the channel noise.

The solution is spectral companding. A spectral compressor continuously monitors the frequency composition of the audio signal and varies the amount of pre-emphasis accordingly. When small amounts of high frequency signal is present, the spectral compressor provides high amounts of pre-emphasis. When high frequency components are strong, it actually contributes de-emphasis in order to prevent high-frequency overload. The end result is that the system frequency response is dynamically adjusted and the transmitted signal consistently contains a substantial proportion of high frequency information. This permits the optimal masking of channel noise and results in a signal perceived as being of high fidelity.

Satellite Stereo Television

Presently, there are four different systems used to transmit stereo sound over satellite circuits. Once this stereo has been decoded by either a stereo processor or a satellite receiver with a built-in processor, the composite baseband audio signal, i.e. the raw audio signal, is fed into a home stereo. Note that only a handful of the new stereo-equipped TVs actually have two separate, high fidelity speakers and do not require an accessory stereo amplifier and speakers.

Multiplex Stereo

The multiplex system is similar to that used in regular FM radio broadcasts. The left (L) and right (R) audio channels are combined in a circuit called a matrix that produces the sum and difference signals, L + R and L − R. The difference signal is modulated onto a 38 KHz subcarrier. (In technical jargon the type of modulation used is double sideband suppressed carrier modulation.) The L + R baseband signal, the modulated L − R difference signal and a 19 KHz reference signal

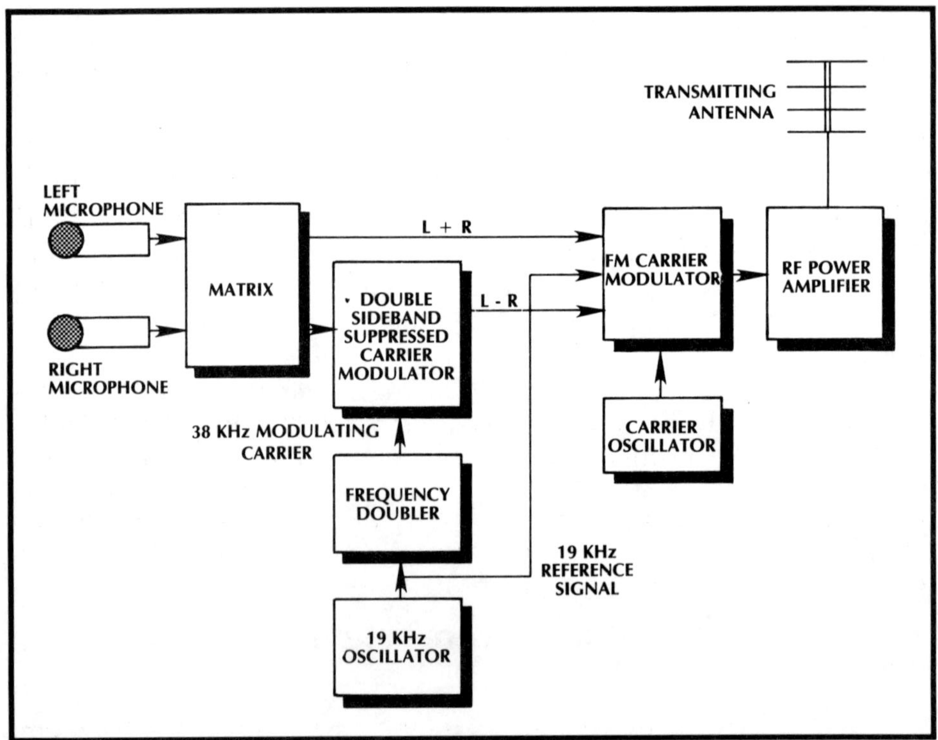

Figure 3-20. Multiplex Stereo. *A multiplex stereo circuit processes the left and right audio channels so that both can be relayed on a single audio subcarrier and then reconstructed at the receiver. The signals are processed into sum and difference components. A 19 KHz reference signal is used at the receiver to adequately separate the two stereo channels. Different forms of multiplex stereo use companding and emphasis to reduce perceived background noise.*

are combined or multiplexed together onto the audio subcarrier. Satellite circuits often use audio subcarriers at either 6.2 or 6.8 MHz.

The 19 KHz reference signal is used to reconstruct the individual L and R components with the proper relative phase from the sum and difference signals. If the phase relationship is garbled, separation between two stereo channels will suffer. This need for a reference signal is similar to the use of the color burst in constructing the three basic color signals from the sum and difference components. In the case of color television, incorrect phase relationship means distorted hues.

The reference signal serves a second function in FM radio. It is included so that the receiver can identify the incoming transmission as stereo audio. FM radio is often used for supersonic audio, audio at frequencies above the human sound threshold of 20 KHz. For example, stock quotations and teletype may be transmitted in this high frequency range. Nothing intelligible would result if an FM radio were to try to decode such signals as audio programs.

Warner Amex Stereo

This system transmits the sum and difference signals over two separate subcarriers. The L + R signal is frequency modulated onto a 5.8 MHz subcarrier; the L − R signal onto either a 6.62 or 6.8 MHz subcarrier. Any combination of these subcarriers or others located in the 5.0 to 8.5 MHz range can also be used. When signals are demodulated at the satellite receiver, they are combined or matrixed to recover the original left and right channel information.

Discrete Stereo

This is the simplest and most widely used satellite stereo system. The left and right channels are not matrixed but each is simply relayed on a separate subcarrier in the 5.0 to 8.5 MHz frequency range.

Processed Narrow Deviation Stereo

Processed narrow deviation stereo is another form of discrete transmission except that very narrow deviation modulation is used. Deviation is a measure of how much the carrier wave is varied from its center frequency. For example, a frequency modulated 10 KHz carrier which varied between 9 and 11 KHz deviates less than the same carrier varied between 7 and 13 KHz. There is a trade-off involved here. Transmissions having more narrow deviations and which occupy less frequency spectrum unavoidably contain greater amounts of noise. Therefore, companding techniques are used.

There are two principle forms of processed narrow deviation stereo. The Wegener system uses spectral companding while the Leaming also employs a second stage of amplitude companding. This results in an apparent 15 dB improvement in signal-to-noise ratio. This is in addition to the approximately apparent 18 dB contribution of the first spectral companding stage. Note that this improvement is called apparent because the average signal-to-noise ratio is the same with or without companding. The improvement is in perceived fidelity. The signal is processed or creatively readjusted to match the sensitivity of the human ear.

Processed stereo transmission has one principle advantage over conventional discrete methods. Many more audio subcarriers can be relayed since the deviation is reduced and less bandwidth is used. A satellite receiver must have a narrow band audio filter to receive such audio broadcasts.

Multichannel Television Sound (MTS)

The multichannel television sound (MTS) system was approved in late 1984 by the Broadcast Television System Committee (BTSC) for use with over-the-air television. It is a combination of the Zenith transmission system coupled with dbx noise reduction, one form of companding. The BTSC system permits the simultaneous transmission of a 15 Hertz to 15 KHz stereo audio signal along with a completely separate audio program (SAP) all within the conventional 6 MHz television channel bandwidth. The SAP provides audio diversity allowing, for example, transmission of a second language or a tutorial channel.

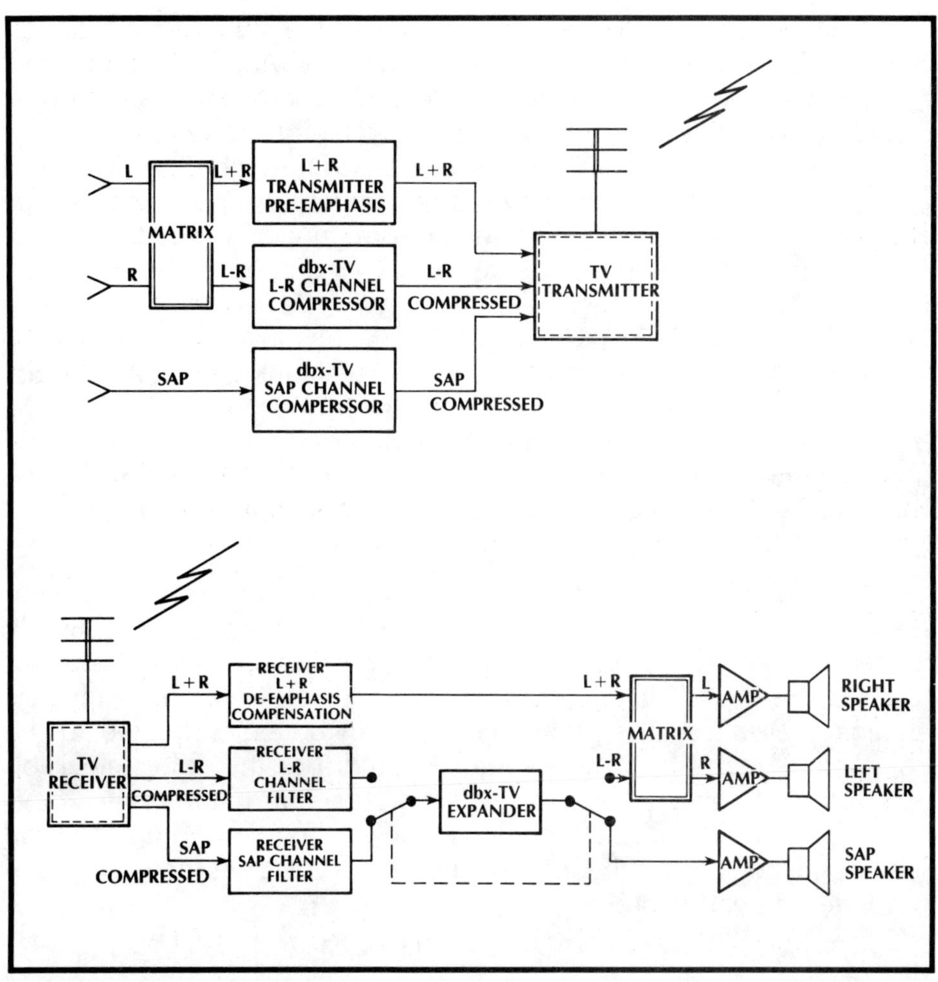

Figure 3-21. MTS Stereo. *The newly introduced stereo format for over-the-air and cable television broadcasts, known as the BTSC format, uses a type of multiplex stereo with a reference carrier frequency equal to the line scanning frequency. The left and right audio signals are processed into a sum and difference channels. The sum signal receives standard 75 microsecond pre-emphasis. The difference and SAP (second audio program) signals are compressed and spectrally companded before transmission. This judicious use of noise reduction techniques results in stereo quality comparable to FM radio broadcasts even when television signals are of marginal quality.*

MTS is designed to be compatible with existing television, just like NTSC color transmissions are compatible with black and white receivers. It is quite similar to the method used to relay stereo over FM radio. The left and right audio channels are converted into sum and difference signals. The L + R replaces the conventional monophonic signal and provides compatibility with non-stereo equipped televisions. When stereo transmission is used, the L − R signal is multiplexed onto a subcarrier having a frequency of twice the reference frequency, which is the line scan frequency. The separate audio program (SAP) is multiplexed onto a subcarrier of five times the reference frequency. Neither the difference signal or the SAP are detected by non-stereo equipped television receivers. Finally, the baseband sum signal and the modulated difference and SAP subcarriers are combined. This signal is used to modulate the standard audio subcarrier within a standard television channel bandwidth.

The reference frequency is none other than the color television scan frequency of 15,734 Hz. (Black and white television has a slightly higher scan frequency of 15,750 Hz.). This value has been chosen for two reasons. First, the need to relay a separate reference signal is eliminated because the audio receiver can be synchronized to the horizontal sync pulse rate. As with FM radio or the color television signal, the reference signal permits accurate reconstuction of the original left and right audio channels. Second, using the multiples of the scan frequency causes the least picture deterioration if receiver tuning is a little off dead center.

Additional processing is required to ensure that the stereo signal is transmitted with as little noise as possible. The baseband sum frequency, L + R, is treated with standard 75 microsecond pre-emphasized at transmission and then equally de-emphasized at reception to reduce noise. Note that both the sum and difference signals receive this type of processing in FM radio but only the sum signal receives standard emphasizing in MTS television. Similar standard emphasis and de-emphasis is used on some satellite circuits but not on MAC transmissions. Since conventional, FM audio receivers in non-stereo televisions contain identical de-emphasis circuitry, the L + R signal is perfectly compatible.

The difference signal is amplitude companded to reduce noise. This is required because the process of subtracting similar left and right audio channels results in a signal which often has a near zero voltage and is "down in the noise" at low as well as high frequencies. Companding boosts the lower voltage portions for transmission and then compresses them at the receiver to restore the original signal. The same companding chip processes the SAP signal.

Spectral companding is also employed to combat the noise problems resulting from the narrow deviation used in BTSC transmissions. Stereo televisions must use a rather limited 25 MHz deviation as mandated by the space allocated to the NTSC audio subcarrier. This is one third the 75 KHz deviation allowed in FM radio. Without use of spectral companding or variable pre-emphasis, noise levels would far exceed those of FM radio which is not a particularly high fidelity audio source.

Receiving Stereo on Non-Stereo-Ready Televisions

Prior to the introduction of MTS, the only method to deliver over-the-air stereo broadcasts was via a "simulcast." The stereo soundtrack was transmitted via an FM radio station to stereo receivers at the same time that the television was viewed. Although most new TV sets are now capable of reproducing stereo sound, the vast majority are not so equipped.

This situation is similar to that encountered in the earlier days of the home satellite market when few satellite receivers had built-in stereo capability. The solution then was the addition of a separate stereo processor for retrofit with older model satellite receivers. These decoders have outputs required to drive the left and right speakers of a standard amplifier.

There are three principle types of MTS decoders available. The dedicated, plug-in decoder can be used so long as the TV has a matching jack and provided the decoder purchased is the same brand as the television. A second type on the market is the tuner-style decoder which incorporates a tuner as well as BTSC decoding circuits. The disadvantage of this variety is that when channels are selected from the TV set, the tuner audio must be changed whenever television

Figure 3-23. Recoton MTS Television Stereo Decoder. *This device uses a small, puck-like probe which attaches to the exterior bottom or rear of any non-stereo equipped TV set and eavesdrops on the RF signal in the tuner. In those few cases where the internal circuitry is too well shielded, the probe must be installed in an internal position. A standard MPX jack output is also provided for those newer model televisions which have such output connectors. The Recoton MTS decoder is designed to provide dynamic, full frequency stereo sound. A spatial expander circuit allows variation of the apparent separation between the stereo speakers. This unit also has high fidelity multiband stereo synthesizer and ambience enhancement circuit with dynamic noise reduction for those cases where only monophonic sound is available. (Courtesy of Recoton Corporation, Long Island City, NY)*

channels are changed. Some more costly hi-fi VCRs also feature built-in MTS decoders and television tuners. A third type is the universal decoder. These feature probes which attach to the exterior wall of a television in order to detect the audio subcarrier. The universal decoders tend to be less expensive than the other two varieties.

Cable Television Stereo

MTS can be transmitted within the standard television bandwidth over cable TV networks. However, this multiplexed form of stereo, like FM radio, is quite susceptible to noise. Care must be taken to ensure that the level of companding is sufficient and that the signal

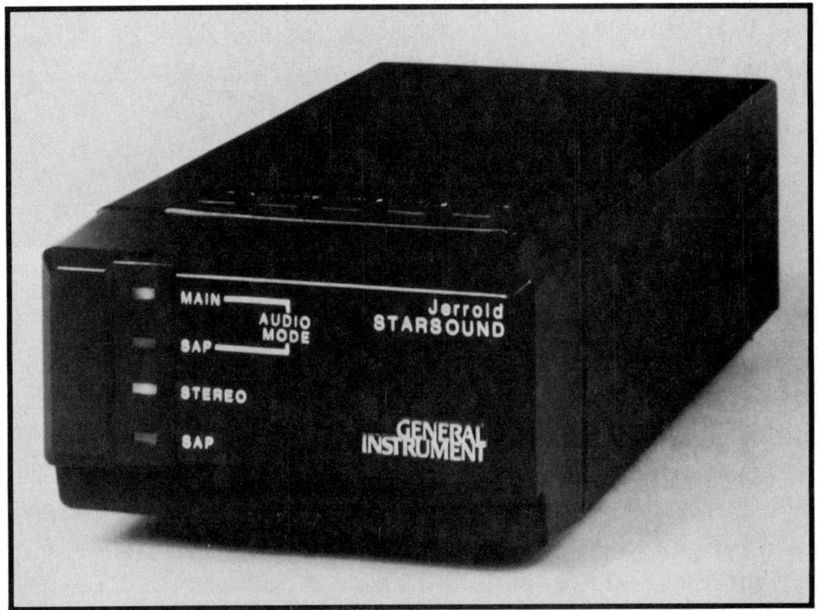

Figure 3-22. Jerrold Starsound Stereo Adaptor. *This low cost BTSC compatible stereo decoder receives and demodulates the BTSC encoded signal relayed over a cable TV network. It can be used as an add-on to the output of decoders and converters or with unscrambled or clear signals to create stereo sound even when using televisions not equipped for stereo. An additional SAP (second audio program) channel is also available for programs such as foreign language broadcasts. (Courtesy of General Instrument)*

level remains strong enough to counteract noise introduced by transmission through long cable runs and line amplifiers.

Stereo can also be relayed in unused video channels including the frequency bands between 108 and 126 MHz and 72 to 76 MHz. Stringent FCC regulations relating to the acceptable amount of signal leakage can be circumvented by transmitting at very low power levels and by using companding to effectively overcome noise limitations.

J. Digital Television

Standard television transmissions are relayed in analog form. Some of the more advanced and innovative methods such as B-MAC, which is discussed in later chapters, are also analog. Even those advanced encryption systems which are classified as digital and use digital processing, in the final analysis, use analog transmission.

The reason that digital transmissions, which are technically feasible today, are not used is a matter of cost and bandwidth. To express a color picture in digital form requires a data rate of between 20 to 80 million bits per second depending upon the coding technique. At the upper limit, this translates into the need for a bandwidth spanning five NTSC channels. In addition, costly high speed, digital-to-analog converters must be used at the receiving end to create a signal which can drive the "analog" television scanning circuits.

In contrast, digitizing an audio signal can now be accomplished at a reasonable cost. A rate of between 200 to 700 thousand bits per second is required and transmission can be easily confined within the standard 6 MHz NTSC bandwidth. Digitized audio is a central feature of advanced satellite television encryption systems.

K. TELEVISION TEST SIGNALS

Television transmissions are accompanied by test signals generated at the broadcast studio. The two most common are the vertical interval test signal (VITS) relayed on lines 7 and 18 in the vertical blanking interval, and the vertical interval (color) reference signal (VIRS) on line 19. They may occupy one or both fields of any television broadcast although neither is mandatory.

These and other types of test signals serve two purposes. They enable a technician to perform fine adjustments to the receiving equipment and indicate the quality of both the transmission link and receiving equipment. Such test signals are transmitted in the television signal and can also be generated at the test bench by devices known as signal generators and viewed on waveform monitors.

Five parameters determine how well a television signal is reproduced. These are: amplitude versus frequency linearity; phase versus frequency response characteristics; transient response; differential gain; and differential phase. Non-ideal reproduction of each of these causes characteristic patterns of distortion in the television picture.

Some of the test signals are illustrated and explained in the accompanying figures. The intent here is not to explore this subject in great technical detail. Interested readers can refer to the reference below or to other books for more detail.

(1) *Television Electronics: Theory and Servicing* by Milton S. Kiver and Milton Kaufman. Delmar Publishers, 1983

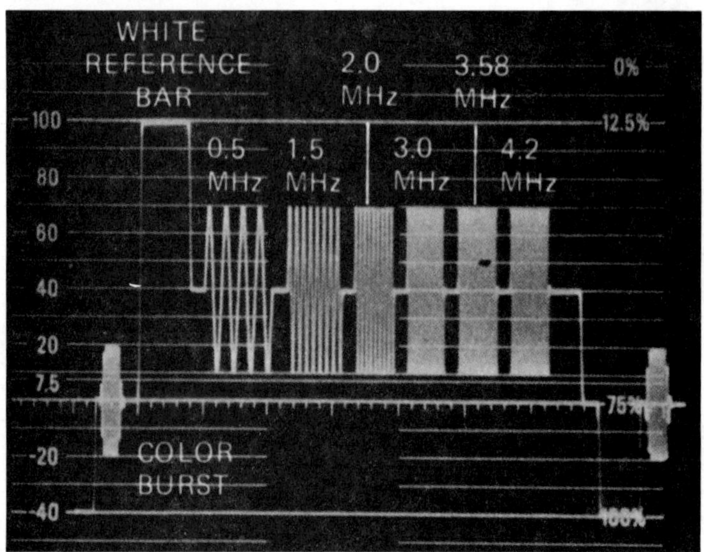

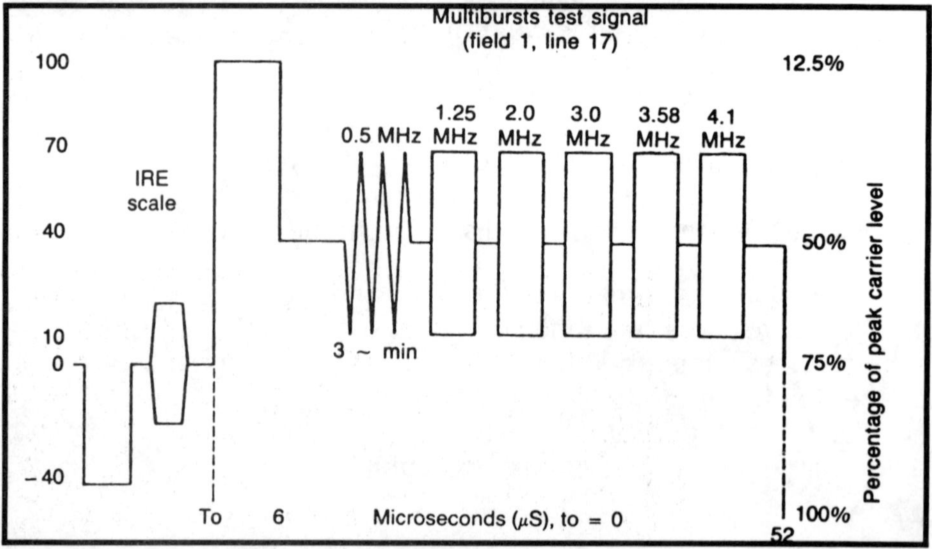

Figure 3-24. The Vertical Interval Test Signal (VITS). *This portion of the VITS usually transmitted on line 17, field 1, is composed of a series of equal amplitude bursts covering a range of frequencies at typically 0.5, 1.5, 2.0, 3.0, 3.58 and 4.2 MHz. If amplitude varies across the frequency spectrum, this will been seen in the height of the bursts. A burst of peak white, known as the "white flag," is also transmitted as a white reference level. Note that the sync pulse followed by the color burst is seen in the lower left hand corner of this trace.*

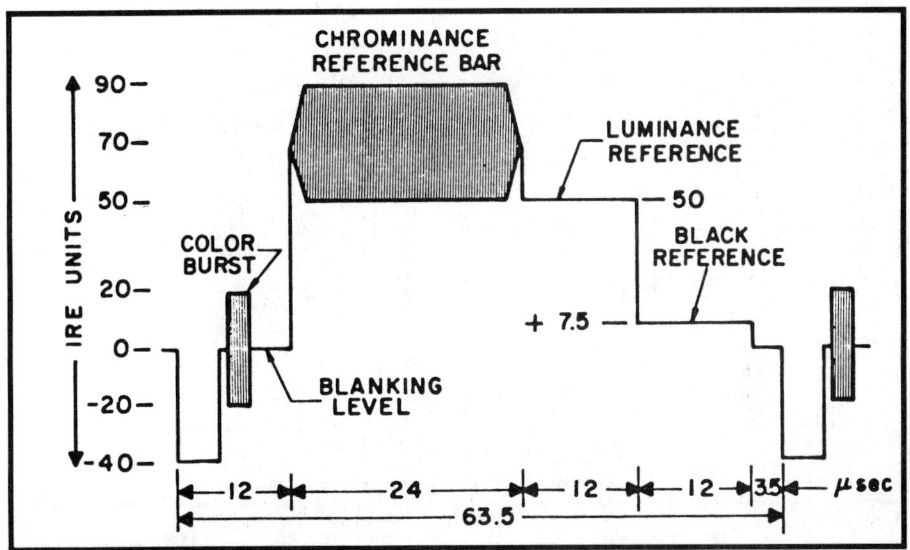

Figure 3-25. The Vertical Interval (Color) Reference Signal (VIRS). *This test signal usually occupies both fields of line 19 of the vertical blanking interval. This signal consists of a chrominance reference waveform at the color burst frequency as well as both a black and white reference level signal. This indicates when the chroma reference and color burst signals are either out of phase or reduced significantly in amplitude, and picture distortion is visible evidence of these changes in phase and color levels.*

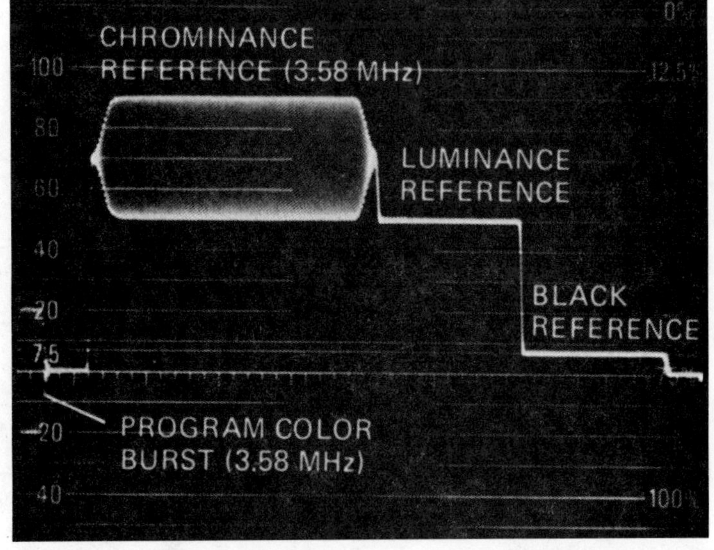

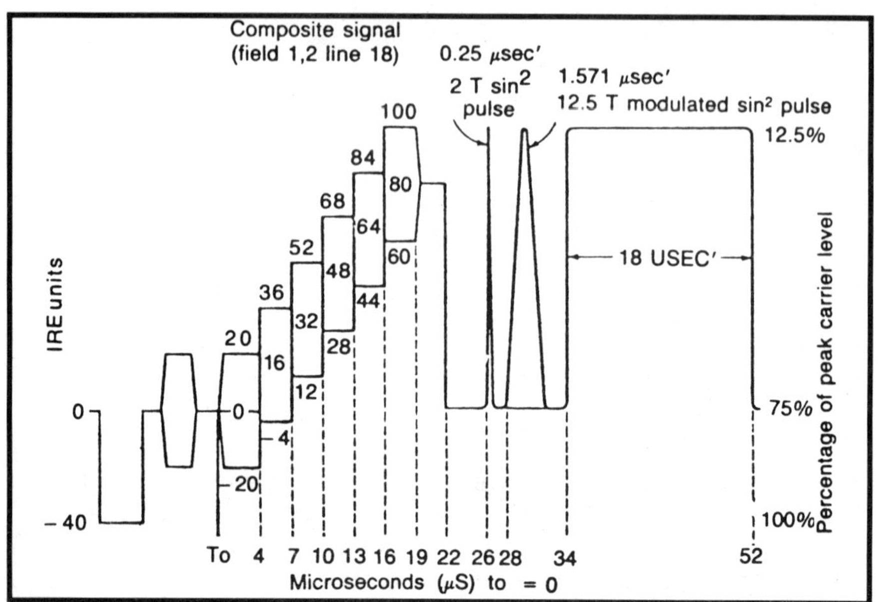

Figure 3-26. The Staircase Composite Test Signal. *This test signal is usually transmitted on both fields of line 18 in the vertical blanking interval. It has a number of components. The first square wave is called the line bar or window test signal. Any tilting of the top portion indicates poor low frequency response visible as picture streaking. The spike which follows, known as the sine squared pulse, is a good indicator of phase distortion. The next wider pulse is referred to as the chrominance pulse test signal and provides an accurate method to determine gain and delay differences between the chroma and luminance signals. The final staircase can be used to measure the amounts of differential gain, or variations in gain across the frequency spectrum.*

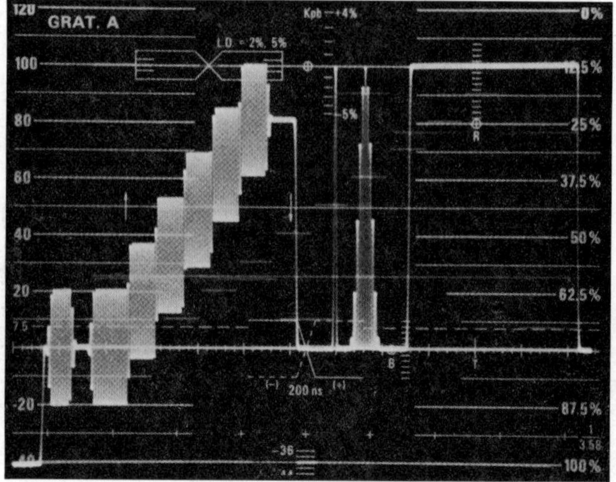

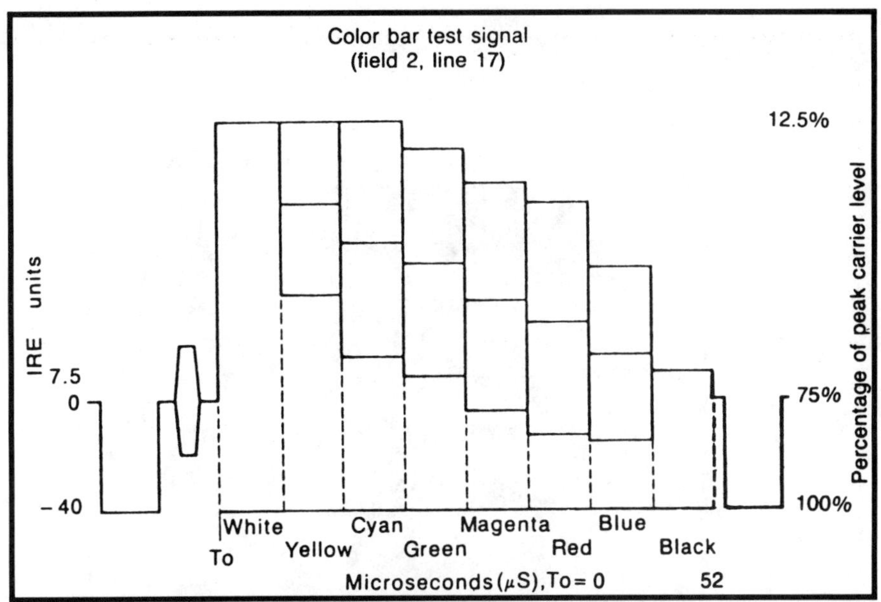

Figure 3-27. The Color Bar Test Signal. *This test signal is usually relayed on the second field of line 17 in the vertical blanking interval. This signal carries the six standard color bars, yellow, cyan, green, magenta, and blue, as well as black and white reference levels. This signal is useful at the studio for adjusting transmitter outputs.*

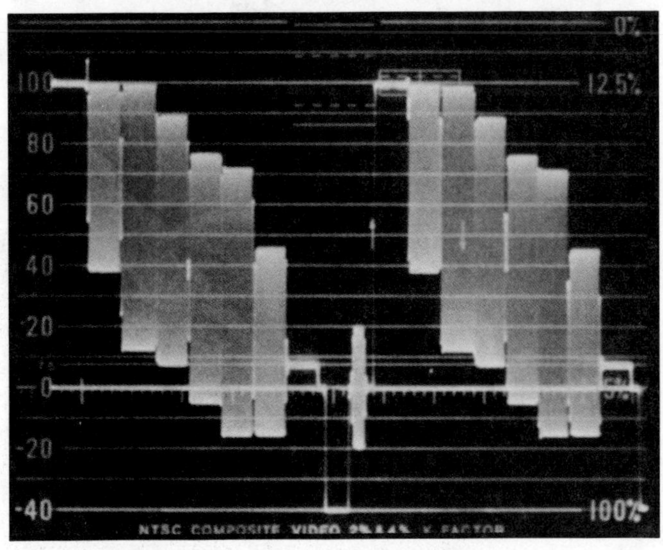

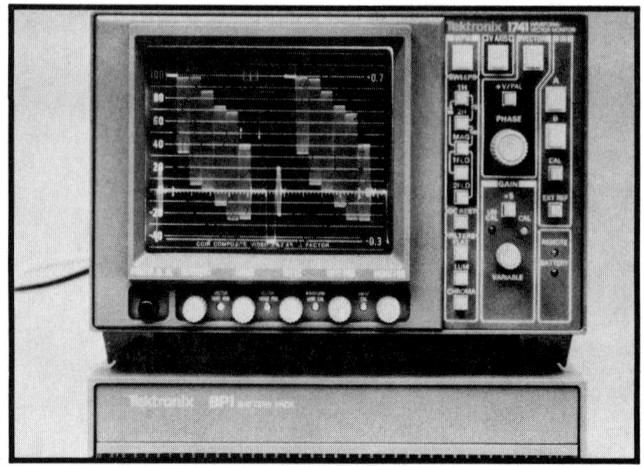

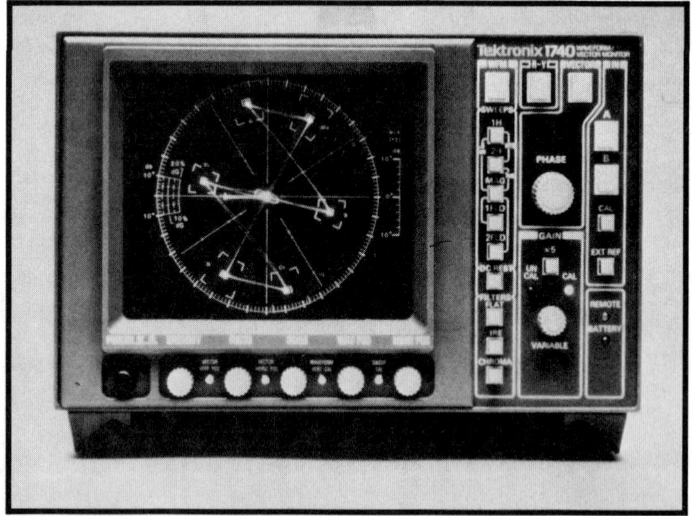

Figure 3-28. Waveform and Vector Monitors. *The 1740 Tektronix Waveform Monitor is capable of displaying all the basic waveform monitoring and vectorscope functions in a single, compact package. Two lines of the color bar test signal are shown in the upper photograph. The bottom photo displays the phase relationships between components of the color signal. (Courtesy of Tektronix)*

IV. SCRAMBLING TECHNIQUES

A. GENERAL CONCEPTS

Television signals are composed of audio and video components. Broadcasts can be scrambled or rendered unusable if either or both of these are altered so that a television receiver cannot recognize the format and cannot reproduce the original material.

Level of Security

The level of security is precisely the degree of difficulty an unauthorized person would experience in recovering the original television signal. The least secure methods can usually be decoded by a competent do-it-yourselfer; the most secure systems generally employ top secret military technologies which are extremely sophisticated and very safe from unauthorized use. It is interesting to realize that commercially available systems use methods that have already been declassified by the military and are not on the cutting edge of technical developments.

The defense establishment will not declassify an encryption system so that it can be sold to the general public unless its code can be broken by their in-house experts. Although military-derived scrambling methods have been the basis of most of today's commercial systems, the heightened interest in protecting broadcasts from piracy has led to the formation of closely guarded industry research groups, and will probably result in development of new technologies independent of their original sources.

Although both the video or audio information are scrambled in most encryption systems, protection against piracy is usually, but not always, established by the level of security of one component of the composite baseband signal. Today, all comsumer-oriented products for receiving satellite Tv broadcasts derive the highest level of security from encrypting the audio information. When lower cost digital video becomes commonplace, this will change. Cable systems have typically established their level of security by the method chosen for video scrambling since audio is often transmitted in the clear.

Four types of information are handled by the most advanced encryption systems: program video; program audio; control; and auxiliary services. More primitive, lower security systems may simply scramble video or a combination of audio and video information.

Video Scrambling

Scrambling of the video information is a process of altering the character of the signal. The video content can be unaffected while non-picture portions of the signal such as the sync pulses are removed, suppressed or shifted from their normal positions. Or the picture information can be altered by inverting its waveform, shifting the voltage level, time shifting or dicing individual lines or adding an interfering signal to mask the entire television signal.

Audio Scrambling

Affordable methods for securely scrambling audio signals are now available because audio bandwidth is relatively more narrow and the rate of information transmission is substantially lower than that of the video portion. Earlier audio scrambling devices used simple techniques such as inserting a secondary subcarrier to hide the audio information, or remodulated the basic audio subcarrier to a higher frequency. Today, however, sufficiently low-cost, high speed, analog-to-digital converters are available to transform the analog audio input to a digital stream which can easily be encrypted to a high level of security. This digital audio signal can then be mixed with control and addressing information and embedded in the video signal for transmission.

Control and Subscriber Authorization

Control information has no direct relationship to the entertainment content of any broadcast. A central question in evaluating and understanding an encryption system is precisely the function of this data stream. Control information can be relayed continuously as an integral component in enabling a descrambler or it may simply be used for one-time subscriber authorization. It can determine whether a decoder can receive just one or a combination of the many available pay-TV channels, known as tiers. Clearly, the overall system design and the interaction between the control data and scrambling methods used determines the overall level of security.

Encryption and Key Distribution

In concept, encrypting a data stream is simple. An input stream is processed according to a formula, known as an algorithm, the specifics of which are determined by the "key." Both the key and the encrypted information are transmitted to a decoder which uses the same algorithm and key to unlock or decipher the data. The end result is an output stream which is the same as the input stream.

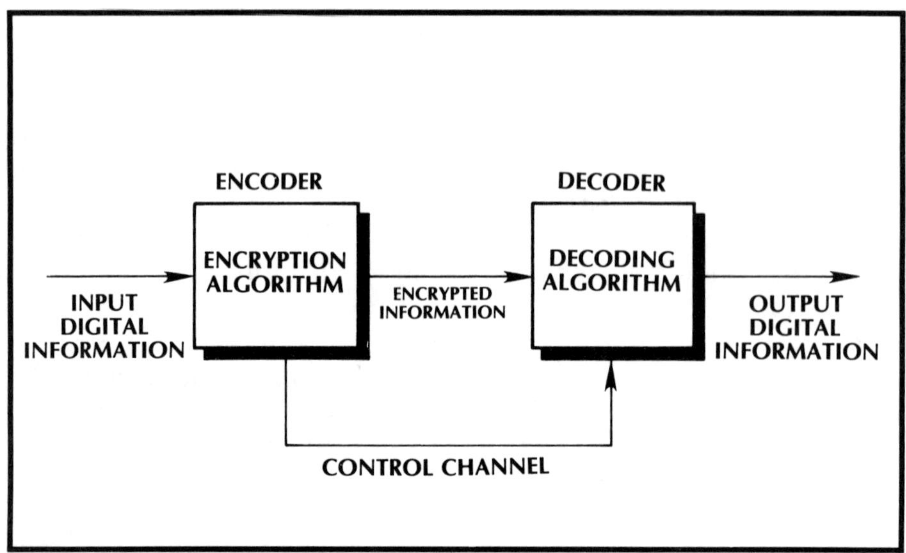

Figure 4-1. Encryption and Key Distribution. *An encryption system which broadcasts to numerous decoders must transmit the required unlock keys with the signal intelligence. Following encryption, keys are relayed in a control channel where they are used to activate the decoding algorithm and to recover the original signal.*

A simple example of an algorithm is instructive. If all the data were expressed as groups of four bits, a choice of an algorithm might be to add the key to the last digit of each group. If the key were 0001, and the original data consisted of five separate groups:

0111 1110 0101 1010 1110

would be transformed by binary addition into a coded form:

1000 1111 0110 1011 1111

This key, 0001, would then be subtracted at the decoder to recapture the original data stream. Since keys are digital words of many bits, many different keys are possible. Two bits can be arranged in 2 x 2 combinations. An 8-bit word can be arranged into 256 possible keys. A 16-bit word creates a total of 65,536 keys.

An entire discipline has evolved around encryption. Since secure transmission of information has been critical for defense and espionage operations, the most highly developed encoding/decoding methods have evolved in secrecy. An ongoing "cat and mouse" game has resulted, especially within the military system, to perfect, break and re-perfect encryption devices.

A "good" algorithm is resistant to being easily broken. No clear and easily detectable relationship will exist between the input and output data or the key. Even if a would-be pirate knew the algorithm, not having access to a key would make deciphering the data difficult, if not impossible. Therefore, a central element in security systems is key distribution or actual security of the keys themselves.

Key Distribution

The system created for key distribution is as critical as creating a secure encryption algorithm. The level of security depends upon the number of bits comprising the key, what channels are used to distribute keys and who may or may not have access to this proprietary information, among other factors. In more advanced systems, the keys themselves are encrypted one or more times with higher level keys. And keys are periodically changed with the knowledge of both the decoder and network control computer. Although keys must be distributed by the broadcast channels when there are many thousands of customers, military encryption unlock codes are actually hand-delivered to their end-users to ensure high security.

It is interesting to realize that scrambling keys to a second level of encryption and therefore creating an additional key does not solve the key distribution problem, it only creates another level of difficulty for pirates. And pirates have been known to be as cunning as those who create the system. In fact, the designers are often the best pirates. The cat and mouse game continues.

Addressing and Tiering

In order to achieve a high level of security, encryption systems incorporate decoders which have a specific address, similar to a telephone number. Individual subscribers or groups of subscribers can be activated or deactivated at will from the central control facility or can be relayed messages regarding upcoming services or information on their billing status.

A global list or customer base is stored in a central computer. Each customer has a serial number and list of authorized services which have been ordered and purchased. A fully addressable system allows this list to be changed at will. This information is relayed to each decoder via the data channel contained within the television signal. Since each customer could be tuned to any one of a number of channels, the information must be transmitted on all the authorized encoded channels or on an out-of-band subcarrier. The control computer cycles through and continuously updates the global list. Each decoder detects all the global data, but acts only on the information transmitted with its unique serial number.

The weak link in a fully addressable broadcast scrambling system lies in its inability to restrict the data flow. Each subscriber receives all the transmitted data so that if an error were made in assigning a key or if an unauthorized address were duplicated, the same set of information would be accessible to all decoders with that serial number. This is a minor weakness in very secure and sophisticated encryption systems.

Addresses are usually embedded in electronic circuits called PROMs (programmable read-only memories) or EEPROMs (electrically erasable programmable read-only memories). EEPROMs can have their memories altered by erasing with an electrical current and then by rewriting new information to memory. In some electronic devices this is accomplished by attaching the appropriate input onto rear panel jacks. Decoders are carefully designed so that the PROM or EEPROM is soldered and mechanically secured to an internal printed circuit board and cannot be easily changed or duplicated.

Today's powerful encryption systems usher in a new "pay-per-view" era. Customers can call 800-numbers to request reception of one-time broadcasts. The decoder is authorized for the duration of the event and is deactivated afterwards. This general ability to allow each customer to select from the total menu of programs available is known as tiering. Ultimately each end-user has the freedom to choose from a menu ranging from short duration pay- per-view fare to programs purchased in monthly or yearly blocks.

The Role of Software in Addressable Systems

The day-to-day operation of an addressable encryption system is controlled at the human/software interface. Even though advanced systems have the potential for being remarkably flexible and powerful, the control computer and its software must be carefully engineered to get the most from the encryption hardware.

This point becomes clear by examining the mechanics involved in addressing customers. An address is created from a combination of zeros and ones. One bit is capable of storing two addresses, either one or zero. Two bits have a capacity of four, two times two. One byte or word which is composed of eight bits has 2 to the eight power, or 256, possibilities. Three bytes or 24 bits can address 16,777,216 subscribers! But three bytes occupy a negligible amount of space in the overall data stream. Therefore, it is the control of this data stream, not the capacity of the data channel, be it the vertical or horizontal blanking intervals or an out-of-band carrier, that is the crucial factor in addressability.

Out-of-Band Versus In-Band Addressing

Numerous methods exist for transmitting the addressing and other data from the control computer to each decoder. Out-of-band subcarriers have been used to a rather limited extent only on cable TV systems. In this type of system the data is modulated onto a carrier embedded somewhere within the entire television channel spectrum. This is usually in the guard band or blank space between VHF channels 4 and 5, or just at the upper edge of the FM radio band at 108 MHz. Each decoder is tuned to this carrier center frequency so that the data stream can be monitored continuously. The primary advantage of this system is that data transmission is independent of the channel being viewed at each subscriber's residence. The disadvantage is that a separate subcarrier is required for broadcasting the data so that extra frequency spectrum space must be allocated. Also, a second receiver is necessary to receive the out-of-band information.

The more common method to transmit data employs an in-band location within each channel's composite video signal. One of two locations is chosen. The 18 to 21 unused lines in the vertical blanking interval are commonly used for establishing the data channel on cable TV systems. However, the data typically broadcasted via satellite systems is embedded within the horizontal synchronization interval. This timing pulse is completely eliminated so that a data stream can be inserted in its place. A matching decoder must then generate the sync pulse as well as remove and decipher the data.

The horizontal blanking interval and its sync pulse may be removed since this information is redundant and can be reconstructed at the decoder from the basic timing information contained in the vertical blanking interval. Each field of 262.5 lines can be packed with a significant amount of data transmitted at sufficiently high rates to allow for addressing of millions of subscribers and for broadcasting all information necessary to dynamically alter the scrambling codes at will. As described in more detail below, the most important non-address information in this interval is digitalized high fidelity stereo sound.

Encrypting Non-Television Transmissions

Non-video portions of the television signal can also be used to transmit information such as teletext, captions for the hearing impaired, on-screen diagnostic messages or stock market quotations. This can be accomplished by the use of character generators built into the decoders.

All the non-video information can be embedded in a shuffled fashion within the audio or control data stream and encrypted hand-in-hand. The more flexible scrambling systems can therefore markedly enhance conventional television broadcasts.

B. SCRAMBLING METHODS

Encryption methods have evolved from very simple devices such as traps to much more complex and sophisticated decoders resembling microcomputers. The basic characteristics of the full range of scramblers is explored below. Once these fundamentals are understood, the details of commercially available encryption systems are examined in detail in Chapters V and VI.

Traps

The simplest scrambling device is a trap. As implied by its name, this device is a notch filter which "traps" out a specific range of frequencies or channels while passing all others. There are "negative" and "positive" traps. In a negative trap system, the trap is centered on the video carrier which therefore never reaches the subscriber. Most but not all traps are very narrow band and do not span the entire 6 MHz television bandwidth.

In a positive trap system, the trap removes a modulated interfering carrier which is almost always 2.25 MHz above the video carrier. In a dual interfering carrier system these signals are located at 2.225 and 2.275 MHz above the video carrier center frequency

A trap is installed "in line" via an input female and an output male F-connector. A cable company could prevent a customer from receiving any selected channel by installing the "appropriate" trap at a location outside a subscriber's home where it is, for all practical purposes, tamperproof. "Appropriate" means eliminating those channels which have not been purchased. A trap is usually secured in a locked box called a cable riser or installed in an apartment or condo utility room.

Connecting and disconnecting a trap is time consuming and not cost-effective because a cable installer must make a service call. Nevertheless, some small, local cable companies still use this type of low security device. A pirate can find and physically remove traps or splice into a neighbors drop line.

Addressable Taps

The addressable tap encryption system also selectively "notches" out television signals before they reach a subscriber's residence. Traps at each user location can be addressed by a headend computer and offer the operator the potential for lower operating costs by permitting changes in basic, premium and pay-per-view services without the need for visits to the residence. A single channel or multiple contiguous channels can be notched out and therefore disabled upon command from the headend computer in fractions of a second.

Addressable taps are designed to provide low-cost addressability for the small cable TV network or a condominium or apartment based SMATV (satellite master antenna TV) system. Traps can be securely installed either indoors in multiunit dwellings or within outdoor cable risers.

Video Inversion

A second simple, low security method to encode television signals is video inversion. The voltage pattern of the video signal is simply reversed so that a television receiver is incapable of interpreting the message. Reinverting the signal, for example, by using a differential output amplifier in a satellite receiver, is the simple decoding necessary. Most satellite receivers on the market today have a rear panel video inversion switch required to decode some of the earlier scrambled signals.

Dynamic video inversion is occasionally used in conjunction with other basic video encryption methods. In this scheme, the video is randomly inverted from one line or frame to the next. A circuit element known as a gate selects between normal or inverted video upon command by the control data, which is required to decode the signal. Dynamic video inversion systems most always are designed to not affect the sync pulses.

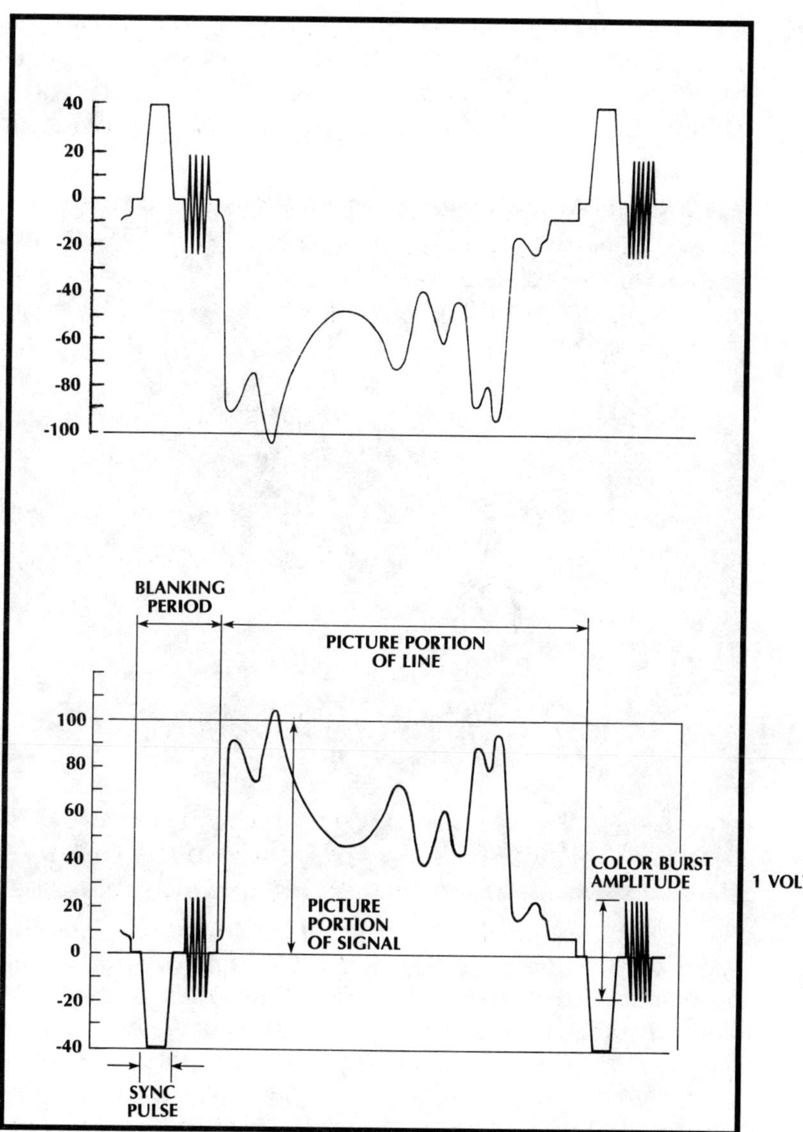

Figure 4-2. Video Inversion. *The entire video waveform can be inverted so that the sync pulses are unusable and the picture information is reversed. This type of video inversion flips the color burst and therefore shifts the waveform 180 degrees in phase. Note that dynamic video inversion usually leaves the horizontal interval in its original form.*

Inverting a signal also affects color synchronization since the color burst is turned upside down. This causes a 180 degree phase shift between the normal and regular color burst and causes the receiver to have difficulties in reconstructing the three raw color signals.

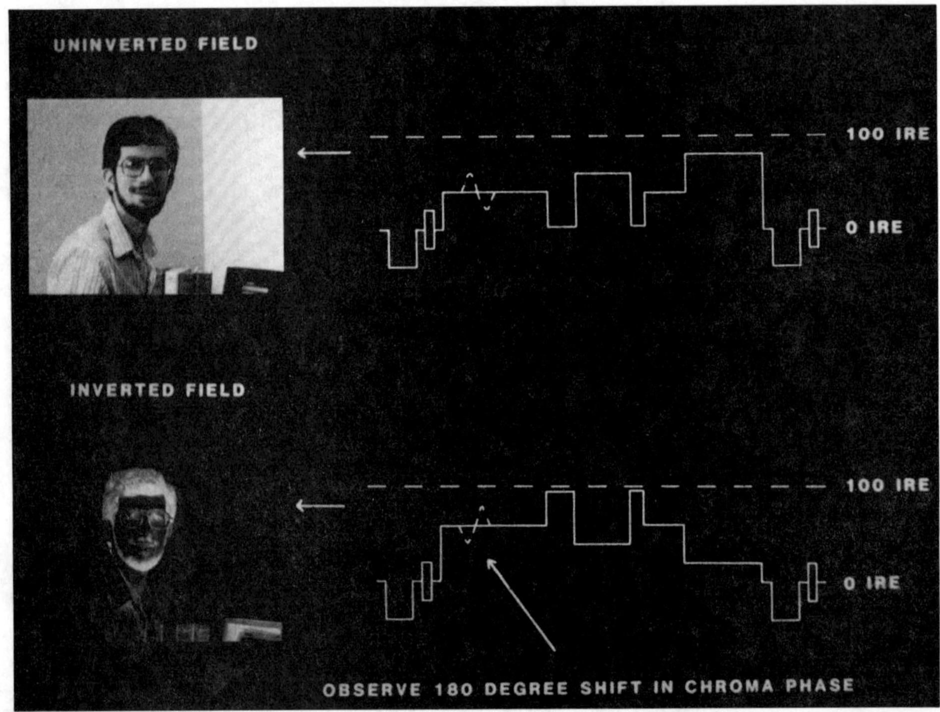

Figure 4-3. Example of Dynamic Video Inversion. *The example shows the effect of inverting the video signal. One line of positive phase video signal accompanies the photograph. The phase of the color burst is inverted and shifted by 180 degrees. This representation of the picture portion of the video signal is an extremely simplified waveform not coincident with that of the photograph. Note that in dynamic video inversion generally only the picture portion of the composite video signal is inverted, not the information in the blanking interval. (Courtesy of Zenith Electronics Corporation)*

Interfering Carrier

A somewhat more sophisticated but still relatively low security approach is to add an interfering carrier at a frequency within the television channel. The narrow bandwidth noise spike is inserted into the frequency space between the video and audio carriers before the video signal is modulated onto its carrier. The signal is physically inserted into the IF loop on a commercial modulator. Noticeable effects of this interference begin at about 40 decibels below the video carrier level with picture "wipeout" occuring when the scrambling signal equals the video carrier level. Therefore, for maximum effect, a voltage near the wipeout level is usually added.

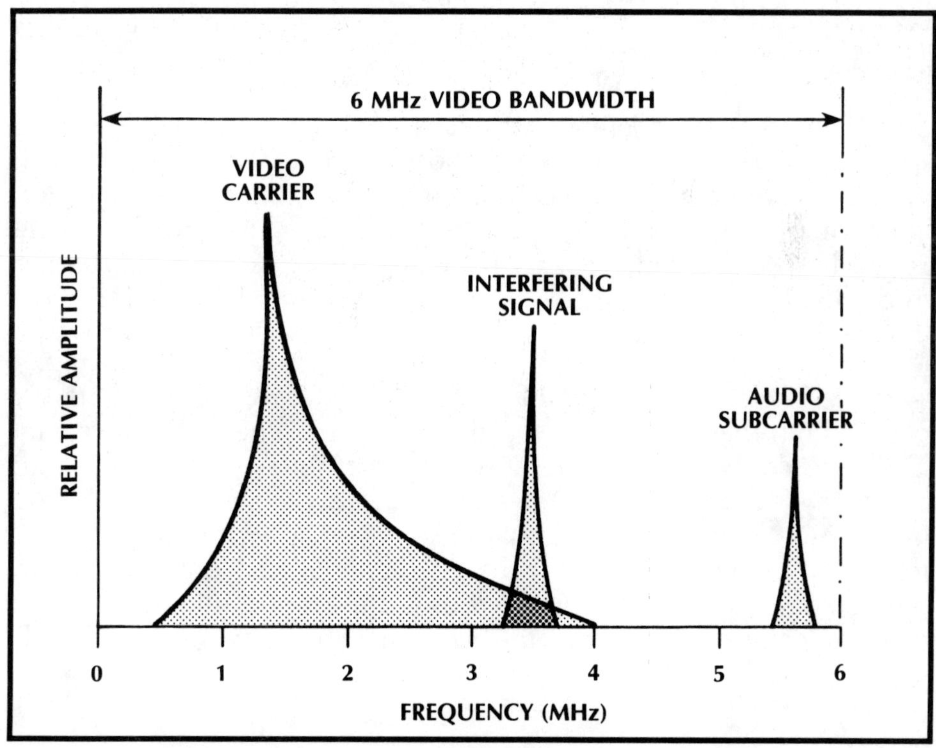

Figure 4-4. Interfering Carrier. *An interfering signal of sufficient magnitude located between the video and audio subcarriers can cause loss of both sound and picture information. This interfering carrier might be a pure sine wave or a signal contained in a band of frequencies. The effect is similar.*

The amount of distortion generated also depends upon the frequency of the interfering signal. Both the video and audio signals will become more impaired as the separation in frequency between the interfering and the signal carrier decreases. Scrambling frequencies which are multiples, otherwise known as harmonics, of the horizontal scanning frequency cause significantly more damage than other interfering signals. The telltale sign of this type of interfering carrier, typically a 1 KHz signal, is horizontal lines which appear across the screen. Such a signal also causes a "tweety bird" effect on the audio output. Harmonics from a 15 KHz interfering signal affect the vertical scanning and cause the picture to roll and jump. Note that the line scan frequency of a b/w television is 15,750 per second which equals 15.75 KHz. These frequencies also impair the automatic gain control and color information circuits. Therefore, the majority of such interference systems combine a 1 and 15 KHz signal together to modulate the interfering carrier.

The successful operation of this type of scrambling system relies on the fact that certain frequency bands within the video signal contain very little information which can be removed without significant picture impairment. This "low impact" frequency is typically between 1.5 and 2.3 MHz above the video carrier center frequency. As described earlier, most always traps are at 2.25 MHz or 2.225 and 2.275 MHz above the video carrier.

The primary advantage of using an interfering carrier is the low cost of both the scrambling and decoding equipment. However, the level of security is also rather low. And a competent do-it-yourselfer can easily build a simple notch filter to defeat the scrambling. In fact, in those areas where broadcasts are relayed with a relatively weak interfering carrier and where there is no channel adjacent to the scrambled channel, detuning the television set away from the interference can be effective in recapturing the black and white portion of the signal.

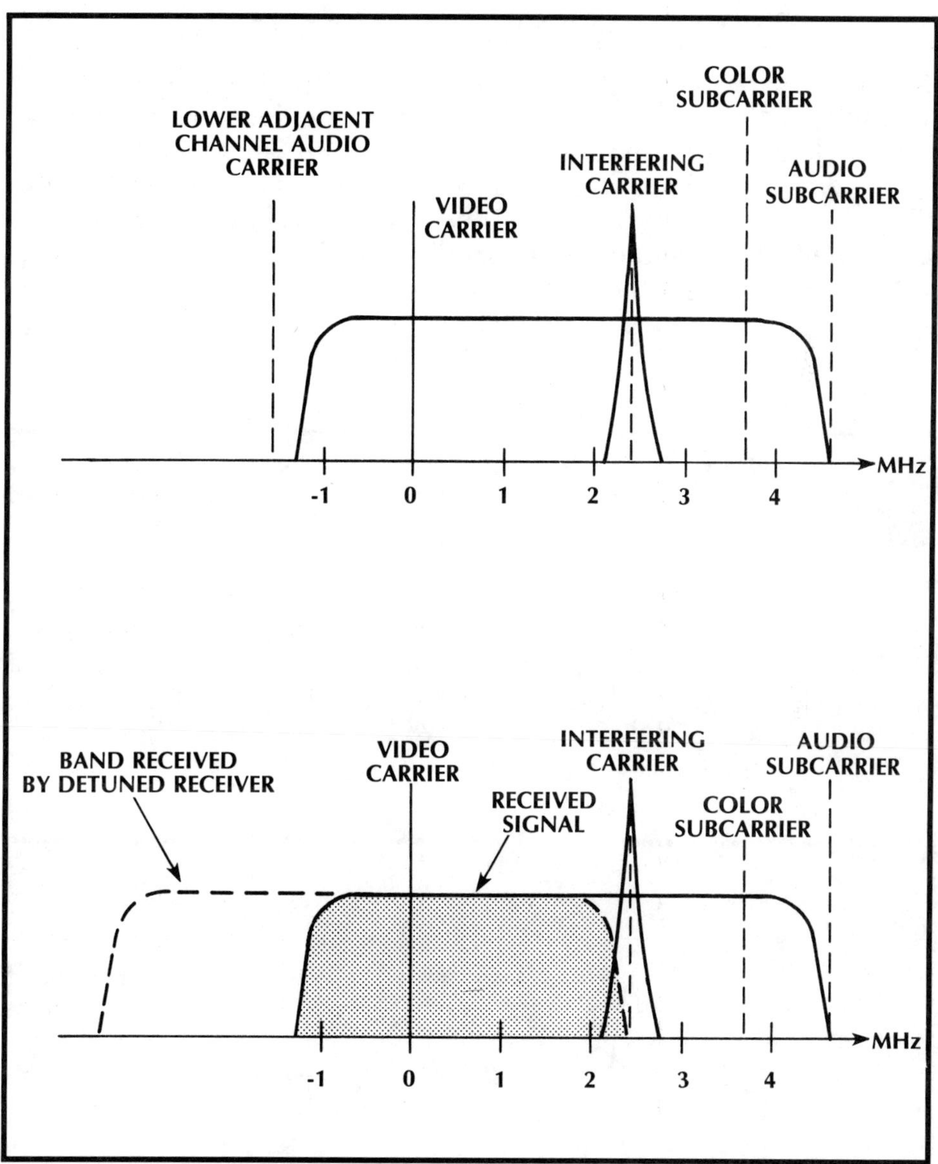

Figure 4-5. Detuning a Television to Recover a Scrambled Video Signal. *Detuning a television away from an interfering carrier can result in recovery of the picture information but eliminates reception of the audio subcarrier. This is only possible if no lower adjacent channel is present. Note that some televisions which are electronically tuned are not capable of being detuned.*

Horizontal Sync Suppression and Shifting

Sync suppression is the most common medium security system that has long been used for scrambling signals on North American cable TV systems. Encoding and decoding equipment is relatively simple and reasonably priced. This method can be combined with a decoder which has addressing capability in order to provide a network operator with a desirable level of flexibility.

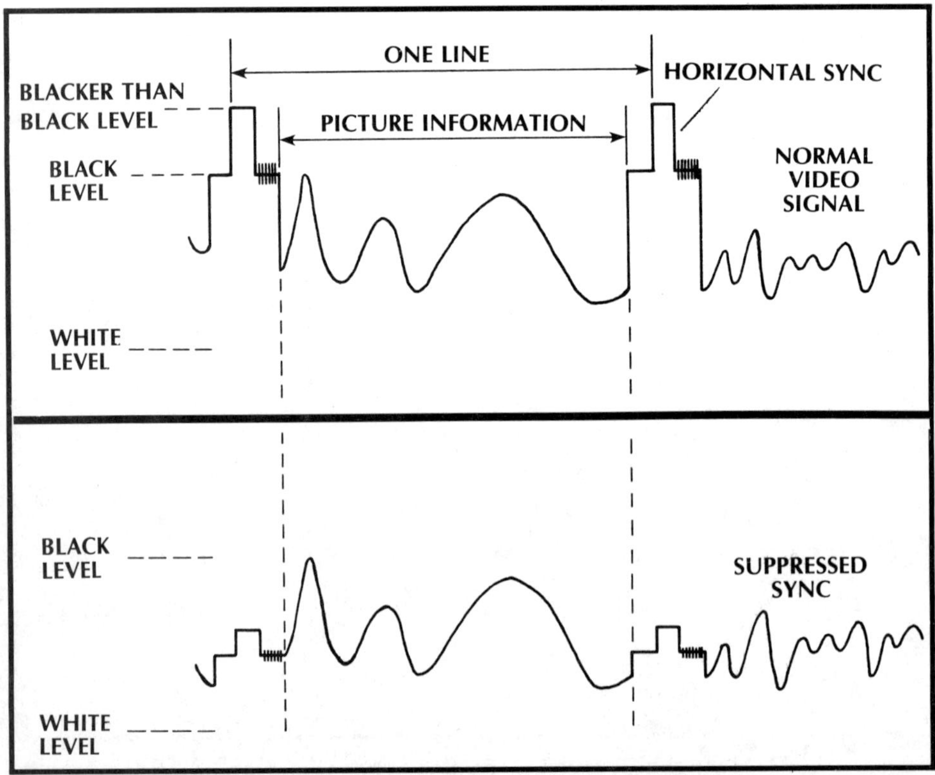

Figure 4-6. Sync Suppression. *Suppressing the horizontal blanking and sync pulse below the black level causes the retrace to appear on the screen and a loss of scan synchronization. While the picture information is unchanged, each line will start at an arbitrary position.*

When the horizontal sync pulse is suppressed to a level below which the TV receiver loses synchronization, the television picture tears or displays a series of wavy lines. This effect occurs since each scan line starts at a somewhat random position on the television screen.

In a cable system, sync suppression scramblers operate at the headend in conjunction with the modulator, the device that processes the composite baseband audio and video signal onto the correct channel frequency. The encoding circuitry monitors the composite baseband video signal and produces a series of pulses having the same timing sequence as the original horizontal sync pulses. These fabricated pulses are then added with the original baseband signal to suppress the horizontal sync pulse amplitude. A 6 decibel reduction in voltage is common and effectively lowers the sync voltage below the black reference level to the range of voltages at which picture information in transmitted.

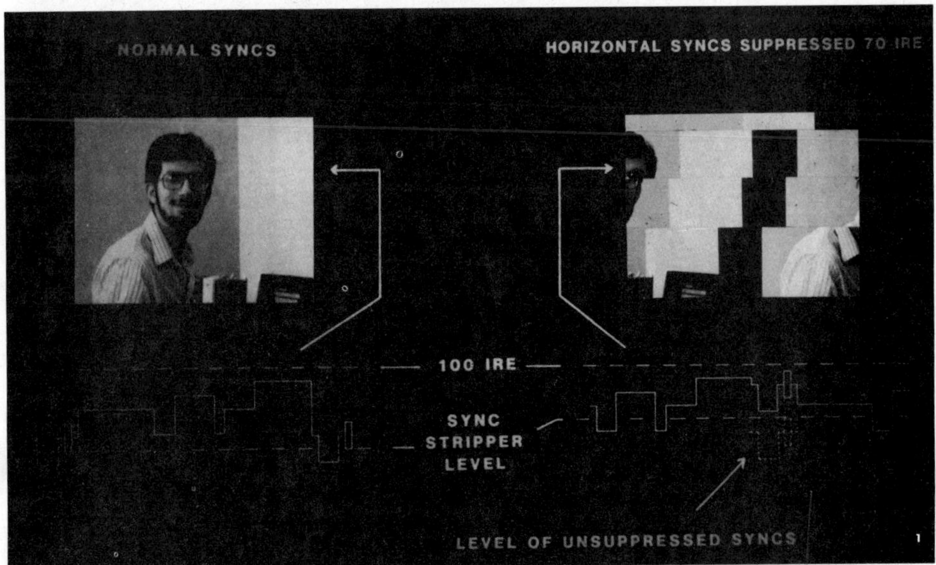

Figure 4-7. Example of Sync Suppression. *Suppressing the sync pulse causes a shift in the line scanning. Here the television picture is broken down into four shifting areas. True-to-life random sync suppression would cause random shifting of each line, not simply picture areas. (Courtesy of Zenith Electronics Corporation)*

Further picture deterioration is caused because suppressing the horizontal sync pulse confuses the automatic gain control (AGC). Television receivers use the horizontal sync pulse as a maximum voltage reference point. This serves in readjusting amplifier gain as the detected broadcast signal varies, so that the voltage levels relaying picture information in each scan line cover the full interval between the reference black and white voltage levels. This ensures that maximum black/white contrast is provided and picture quality remains high. Without the horizontal sync as reference for the AGC, this circuit assumes that the highest voltage levels in the video signal are at the reference black level and, as a consequence, non-black, relatively high voltage level picture regions are mistakenly interpreted as completely black.

Horizontal sync suppression scramblers do not affect pulses in the vertical blanking interval. These remain intact and are used by decoding circuitry to re-establish system synchronization or lock. Some encryption systems have been designed to transmit the necessary decoding pulses on the audio subcarrier. As a result, signal reconstruction is not necessary and less complex decoders can be used.

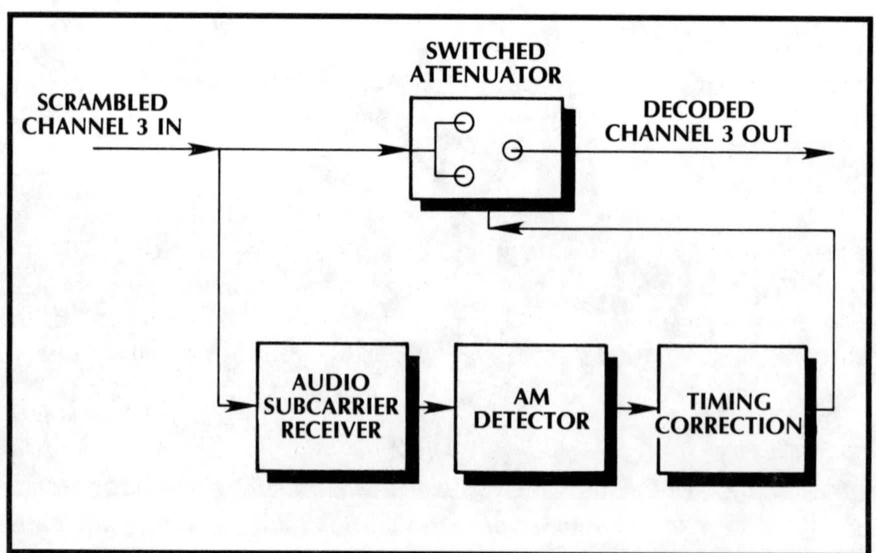

Figure 4-8. Sync Suppression Decoder Block Diagram. *One simple method to recover sync is to extract the timing information often transmitted along with the audio subcarrier.*

Suppressing the horizontal sync pulse has the disadvantage of reducing the video signal-to-noise ratio. This relative reduction in video level occurs in the same ratio as does the sync suppression. This happens since the video signal must be reduced by the same ratio as that by which the sync level is suppressed to ensure adequate reconstruction.

Sync suppression is only a moderate security scrambling system. Cable operators have been plagued by the existence of "black box" or unauthorized decoders. Many television viewers have managed to view the premium programming while avoiding payment of the required monthly subscription fee. In fact, bootleggers have been known to bid head-to-head with larger cable companies at auctions in order to purchase decoders from a bankrupt cable system. The unauthorized units were then internally jumpered to defeat addressability.

Hardware manufacturers have attempted various remedies to curtail signal pirating. The security of the basic sync suppression system has been enhanced by adding multiple levels of suppression so that finding the correct level to reconstruct the sync pulses would be more difficult. The decoding pulses carried on the audio subcarrier have also been shifted in time in order to increase the difficulty and complexity of deciphering. In some of the earlier systems the addressing information was carried out-of-band on a separate subcarrier. This data was therefore less easily detected.

Most sync suppression decoders operate at the channel 3 or 4 frequency and are intended to be used in conjunction only with standard cable TV converters. In fact, some brands do not work properly unless mated with a matching converter. Other varieties of descramblers have been integrated into a standard cable converter.

Television decoders that do not operate at baseband frequencies are designed to convert the signal onto channels 3 or 4. This allows common decoder circuitry to be used for any selected channel.

Sync Removal

Complete elimination of the horizontal sync pulse is a more secure method than either suppressing or shifting this timing information. The key to decoding then lies in reconstructing the sync pulses and inserting them at precisely the correct location onto the composite baseband video signal. There are a number of ways to accomplish this feat.

Integrated circuits are available which generate the full set of sync pulses required in the television signal. These devices can be synchronized to the incoming signal by detecting the built-in clock signal which is transmitted with the data, and setting the trigger at the appropriate point for lock up. The sync generator runs continously and its output is correctly added onto the video signal during the horizontal sync interval. A DC voltage reference obtained from pulses in the vertical blanking interval is used to set the reconstructed horizontal sync pulses to the correct level.

Sync pulses can also be generated by using a PROM and a resettable electronc clock. The PROM can generate a voltage pattern which resembles the sync pulses. Voltage levels are established digitally and the clock speed is chosen so that the smallest necessary time increments can be resolved. Once the sync starting point has been established, the clock is started and the output of the PROM is fed to circuitry which gates or injects it at the appropriate moments into the video signal. This pattern replaces the original sync pulses with new ones precisely matched in timing and voltage levels.

Baseband Addressable Systems

Baseband addressable encryption systems were developed to overcome the limitations of sync suppression methods. This new generation of scrambling equipment performs encoding and decoding operations on the baseband video signal, not at the intermediate (IF) or radio (RF) frequencies.

The design objective was to provide improved signal security at a reasonable cost. The pioneer system, introduced by Zenith, uses a multilevel scrambling method which is described in more detail in the following chapter.

Line Dicing

Line dicing is a high security scrambling technique used with digital video encryption. Portions of any scan line can be removed and spliced onto other lines according to an encryption algorithm. Therefore, for example, the first 20 microseconds of the total 63.5 microseconds of line 10 could be removed and added onto the end of line 17. Or lines can be divided into segments which are interchanged in position for transmission. Portions of the line can be repeated to mask the effect of this "splice." This type of scrambling is expensive since it requires the use of digital video processing.

Digital Audio Encryption

Encryption of satellite broadcasts must necessarily be very secure considering the enormous audience and the equally large potential for pirating. The systems chosen to accomplish this task rely on digital scrambling of both the audio and data. The video can also be scrambled at a lower level of security by removing the horizontal sync pulse. The digital audio is inserted in its place along with decoding, addressing and miscellaneous data. Some high security systems such as the M/A COM VideoCipher I also use line dicing in addition to digital audio encryption.

A digital signal is captured by the decoder, fed into a temporary storage area called a shift register and then loaded into a microcomputer for processing. It is descrambled by reversing the effect of the coding algorithm. (One such well-known algorithm, used in the VideoCipher II system is the digital encryption standard (DES) which was developed by the National Bureau of Standards.) This unshuffles the digital stream back to its original form so that the data and audio can be separated and reconstructed like sorting mail into two separate stacks by zip code. Then the audio is converted to an analog signal for amplification and reproduction.

The description or unlocking key is also transmitted along with the data stream and must be separated to activate the decoder and reverse the effects of scrambling. Note that this key does not contain all the deciphering information, a large part of this is hard wired into the descrambler circuitry, but it is merely the catalyst that unlocks the power of the decoding circuitry.

The security of many satellite scrambling systems hinges on the security of the audio information because the video scrambling is often relatively soft. An important advantage of this system is that audio reproduction can actually be improved by encoding/decoding since is it digitally relayed. While analog messages are subject to signal-to-noise degradation during transmission, streams of 1s and 0s must only be recognized as either a presence or absence of a pulse to be perfectly reconstructed. Therefore, using a digital format can potentially improve audio transmission.

Since information other than audio is transmitted in the horizontal blanking interval, these systems have great flexibility. Addressing of each subscriber takes place on a dynamic basis because the headend computer can cycle through its customer list and can download authorization information as well as any other message desired. Every terminal in the system receives all of the information, but acts only on that data addressed to it. One decoder can be used to unlock any scrambled channel when addressing information with the proper authorization is transmitted from the control computer.

C. IMPROVED TELEVISION TRANSMISSION SYSTEMS

The NTSC color television system and its PAL and SECAM offspring were advanced technological achievements when first developed but today have limitations which are especially evident when viewing satellite TV broadcasts. This set of problems coupled with recent developments in data storage devices and microprocessors based on large scale and very large scale integrated (LSI and VLSI) circuits have served to foster the development of a new broadcast format known as high definition television (HDTV). A first step in this direction is the multiplexed analog component (MAC) system, originally created by the British Independent Broadcasting Authority and now being perfected and marketed by Scientific Atlanta and Plessey.

The MAC system transmits higher quality picture and sound, makes more efficient use of the available bandwidth and lends itself to more secure encryption techniques than conventional television broadcast methods.

Limitations of Existing Broadcast Systems

Experts have claimed that NTSC, PAL and SECAM transmission systems are already obsolete. Numerous technical upgrades and limitations can be identified. One of the principle shortcomings is the fact that up to 20% of the total picture time is dedicated to synchronization pulses. And these sync pulses are relayed at relatively high voltages causing an unnecessary consumption of power.

The overriding consideration in designing satellite circuits is the noise constraint. Detecting satellite signals is a painstaking exercise in recognizing very faint signals in a background of noise because low powers are transmitted over extremely long distances. Neither PAL nor NTSC were designed for use in very noisy environments. These standards were developed so that AM signals could be efficiently broadcasted to local television receivers. However, satellite broadcasts must be frequency modulated.

FM and AM broadcasts have different noise characteristics. AM transmissions have "flat" amounts of noise picked up en route from the television station to the set. This means that the same amount of noise will be detected at 2.1 MHz, 3.6 MHz or any other frequency in the spectrum. The design objective to have both the chrominance and luminance in the television signal adjusted so that they would have equal amplitudes after being processed by the television receiver was easily accomplished. However, FM signals pick up higher amounts of noise as transmission frequency increases. Therefore, the chrominance signal becomes noisier than the luminance component en route to the television and picture distortions can be the outcome. Circuits to correct this effect cause problems of their own, although at a less noticeable level.

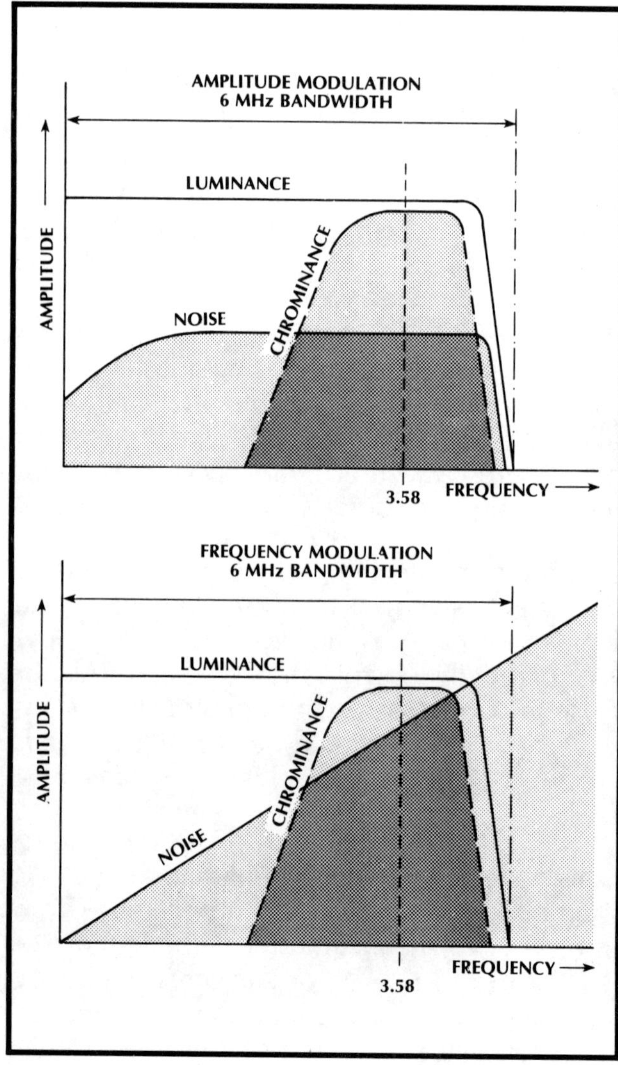

Figure 4-9. FM Transmissions and Noise Problems. *In AM over-the-air television broadcasts the noise is relatively constant over the entire 6 MHz wide television bandwidth. This facilitates balancing the luminance and chrominance signals and avoids picture distortions. By contrast, FM broadcasts are more susceptible to interference from noise at higher frequencies. This lowers the signal-to-noise ratio of the chrominance relative to the luminance component of the total signal and can increase color distortion.*

Conventional television transmission techniques also suffer from the undesirable overlapping of chrominance and luminance signals. This occurs because the AM video and audio signals must be multiplexed together or combined for transmission. Under adverse conditions the luminance gets into the chrominance signal and is interpreted as color information. This effect is known as cross-chrominance and shows up as spurious color bands rippling through regions of repetitive fine detail when images such as a striped shirt are displayed. The opposite effect, known as cross-luminance or cross-color, is noticeable as a travelling dot pattern in regions of saturated colors. Although there are technical fixes to these problems, none are perfect or inexpensive.

Problems also arise when extra audio subcarriers are transmitted via satellite along with the television picture. These use some of the available power and leave less for transmitting the video. When numerous audio subcarriers are relayed, video quality can be noticeably degraded.

The NTSC and other formats also do not make use of the available bandwidth as efficiently as possible. Wider bandwidths are required only when illumination or sound levels have marked changes. However, most often picture information changes slowly from one scan line to the next. Therefore, a method which would code the signal on changes between lines rather than repeat similar information for each scan would use less bandwidth.

High Definition Television

Technical and Performance Objectives

A number of objectives for HDTV can be identified. Higher resolution video with improved clarity is the most desirable feature. This includes enhancing both horizontal and vertical resolution to make large-screen TV viewing more realistic. A related objective is to as much as double the 4 to 3 aspect ratio so that television pictures would more closely match those seen on large screens in movie theaters.

HDTV should also minimize and preferably eliminate cross-color and cross-luminance effects as well as the restriction on the luminance signal bandwidth as additional audio subcarriers are added. Finally, those picture imperfections categorized as aliasing, namely the flicker or motion strobing and line visibility, should not be noticeable.

HDTV should also have improved audio performance as well as adequate space to relay multiple audio channels. Digital audio is a preferred format to achieve these objectives. Additional audio channels could be used to transmit foreign languages or captions for the hearing impaired in addition to providing high fidelity sound. Present systems have limited space in the vertical and horizontal blanking intervals for insertion of services such as teletext or high capacity data channels.

Television viewing would also be enhanced if the available bandwidth were used more efficiently. FM satellite broadcasts should also have balanced chrominance and luminance components in order to improve picture fidelity.

Finally, a well-designed HDTV system should be easily and inexpensively scrambled and decoded to a high level of security.

An immediate benefit of eliminating multiple worldwide broadcast formats would result if a single high definition television standard were adapted. In particular, this would be welcomed in Europe where incompatible PAL and SECAM color schemes are in use.

Present Status

A number of operational high definition television systems have been demonstrated in the United States and Japan. However, these employ digital transmission with almost twice the number of lines per field and twice the screen aspect ratio. Without use of bandwidth conservation techniques, these high quality pictures require as many as five satellite TV bandwidths for transmission. Even when information redundancy and bit-rate reduction methods are used to their fullest extent, at least two full color television bandwidths are required for transmission. The A/D and D/A converters necessary to accomplish this task are presently prohibitively expensive for consumer units.

Sophisticated data compression techniques which can lower the required bandwidth are also costly.

A further difficulty lies in the path of implementing HDTV. Existing television cameras, transmission equipment as well as receivers are not compatible with this new system. A gradual changeover to the new technology could occur if intermediate, expensive converters were available. The present-day reality of the situation has inspired the development of a very effective half-way technology, the MAC system.

The MAC System

Multiplexed analog component transmissions completely separate the audio/data, chrominance and luminance signals and sequentially relay them in compressed form on each scan line. This is a totally different approach to that used by the previous standard broadcast systems which mix or frequency multiplex these three components together. In B-MAC transmissions both the color and luminance signals which are relayed in analog form are expanded by the television receiver and processed so that each scan line begins at the correct starting time. The two difference signal components of the chrominance signal are interlaced. Even lines are transmitted with one color component and odd lines with the other.

MAC transmissions are ideally suited for satellite television. When signals are time compressed they require extra carrier bandwidth in order to relay the same amount of information. FM satellite broadcasts are designed to expand the 6 MHz television bandwidth onto a 36 MHz bandwidth so that the signal-to-noise ratio of the transmitted signal is maximized. More than enough bandwidth is available for MAC transmissions.

Circuitry designed to receive the MAC broadcasts must be capable of storing one or two scan lines. The time-compressed signal must be expanded and resynchronized to the correct starting point. The expansion can be accomplished either using digital storage or charge-coupled devices (CCD). Although the former technology has certain marginal advantages, CCDs benefit from being immediately available and inexpensive.

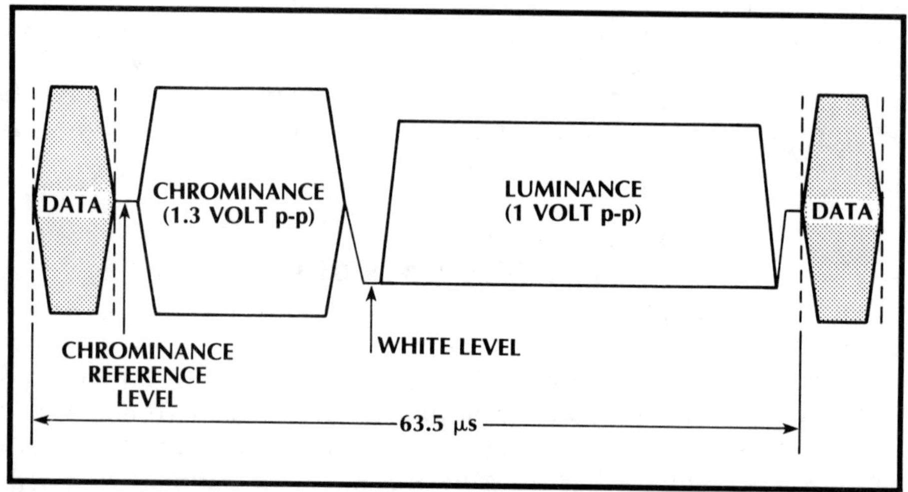

Figure 4-10. Scientific Atlanta MAC Transmission Format. *Data, chrominance and luminance information are transmitted in a compressed form at different times within one horizontal scan line. This technique avoids some of the problems with picture distortion experienced with conventional broadcast formats.*

MAC transmissions have some very attractive advantages over NTSC, PAL and SECAM systems, but the output of MAC encoders are compatible with and can be viewed on conventional televisions. Cross-color and cross-luminance effects are completely eliminated because these components are relayed at separate times within one scan line. This transmission process is known as time division multiplexing. Since the luminance signal is relayed at baseband, color noise is reduced and color bandwidth is increased. The synchronization information requires only 0.2% of the total transmission time compared with over 20% required for conventional systems! The sync is extremely "rugged" meaning that the line trigger point is unlikely to be mistaken by a television receiver. The transmission time which has been freed up can be filled with other digitalized information. Since subcarriers are not used with the MAC system, satellite receivers can be pushed to their low threshold limit. Antennas up to 20% smaller in surface area may be used in comparison to those required for reception of conventionally formatted broadcasts. The electronics have been designed so that the raw red, green and blue video signals are available for improved television display. MAC also has the capability to increase the screen aspect ratio.

Four different MAC systems have been identified. The differences lie principally in the way the data/audio information is transmitted. For example, C-MAC, which is probably the system of choice for future European DBS (direct broadcast systems) television, the RF carrier is time multiplexed during the line blanking interval. Up to eight high fidelity audio channels can be relayed. In B-MAC, now being implemented in Australia and used in Holiday Inn's Hi-Net system, data is time multiplexed at baseband allowing accommodation of six audio channels. The digitally coded audio and data are inserted as bursts of "bits," ones or zeros, in the line blanking portion of the video waveform before being modulated onto the transmission carrier.

TABLE 4-1. MAC AUDIO FORMAT METHODS

	FREQUENCY MULTIPLEXED	TIME MULTIPLEXED
BASEBAND	A-MAC	B-MAC
RF	D-MAC	C-MAC

Audio channels in the MAC format are designed to be compatible with the highest fidelity sound. Transmissions use variable pre-emphasis. Pre-emphasis is a standard form of noise reduction in audio engineering used, for example, in conjunction with all FM broadcasts. Noise in FM transmissions increases with the frequency of the signal. Before the signal is transmitted, the higher frequency portions are pre-emphasized or boosted in power relative to the lower frequency

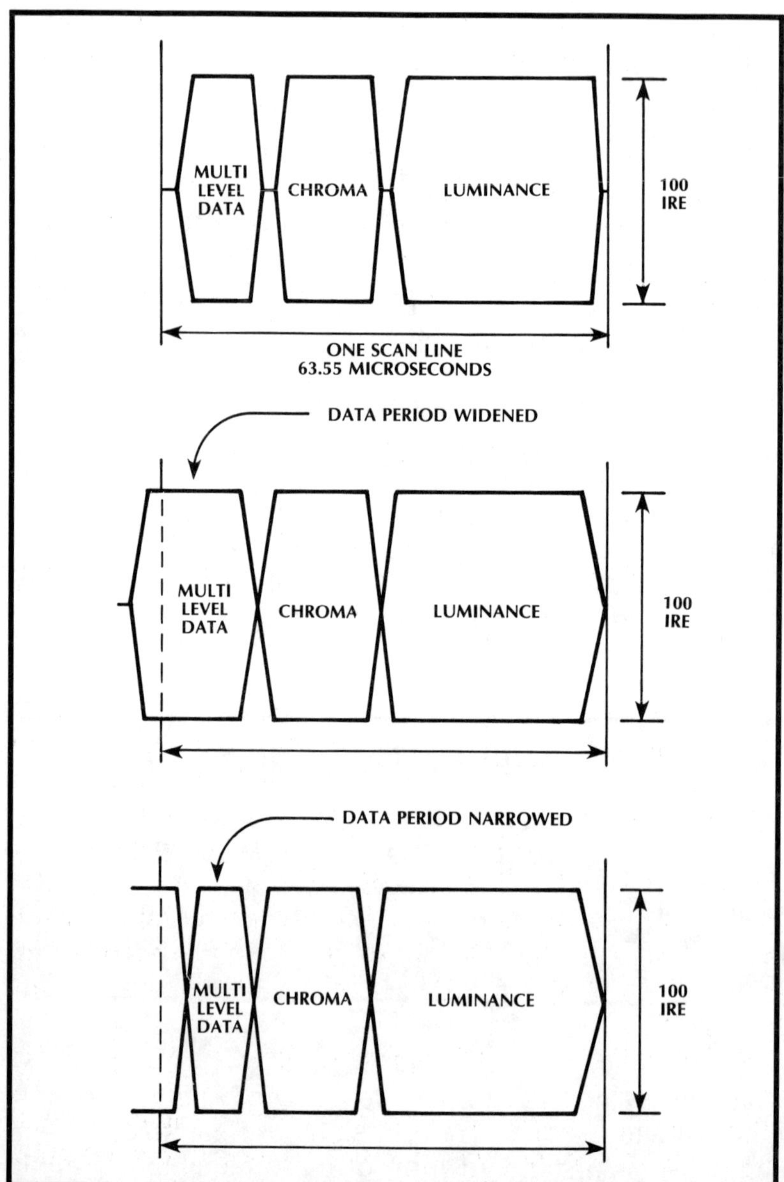

Figure 4-11. MAC Line Shift Scrambling. *Increasing or decreasing the width or duration of the data portion of the line causes a shifting of scan start point and is a very effective scrambling method.*

portions. After reception, the higher frequency segment is attenuated by an equal amount so a balanced signal results. In the process, the extra noise picked up during transmission is reduced and the signal-to-noise ratio is increased.

The amount of pre-emphasis in conventional FM broadcasting is set by international agreement. A better, more complex solution is used in MAC transmissions. The signal is sampled before broadcast and the amount of pre-emphasis is continually adjusted to minimize channel noise. Please see Chapter III for a more detailed description of this form of spectral compounding.

Encryption and Addressing of MAC Transmissions

MAC video can be scrambled in a number of ways including sync removal, video inversion, line shifting or line shuffling. Line shuffling or translation, the preferred choice, is a very secure encryption method available at a reasonable cost. In this system, the video information is shifted in time from a minimum of zero to a maximum of twice the horizontal sync period. This is accomplished by widening or narrowing the data period by a few microseconds to produce shifts in start time of the active video. This time translation is small on each line but the effect is cumulative, resulting in random patterns being shifted by up to one line. The shifting can be pseudo-random. The picture is so totally obscured that simple sync reinsertion creates no noticeable improvement. A high level of video security is accomplished by using the DES algorithm to encode the scrambling key. In short, this encoding technique is attractive because it barely degrades the signal, maintains good video integrity, is relatively inexpensive once the MAC system is in place, and provides relatively high security.

The audio signal on MAC systems is already in digital form so that very secure encryption based on the DES algorithm can be used. During line-translation scrambling, the number of symbols per line is varied pseudo-randomly while the average number of characters per line remains constant at 78. All other digital signals such as text, data, business communications or even addresses can also be encrypted.

Figure 4-12. B-MAC Scrambled Picture. *This illustrates a television picture of a bouquet of flowers before and after scrambling by the B-MAC line shuffling method. (Courtesy of Scientific Atlanta)*

The decoding keys and codes based on a 56-bit word can be changed four times per second.

The data portion of the line can store up to 256 million addresses and access them at a rate of 1 million per hour. Each MAC decoder/receiver has built-in multiple addresses for independent accessibility and control by different programmers. The subscriber can switch between different channels and be instantly connected to any of the programmers.

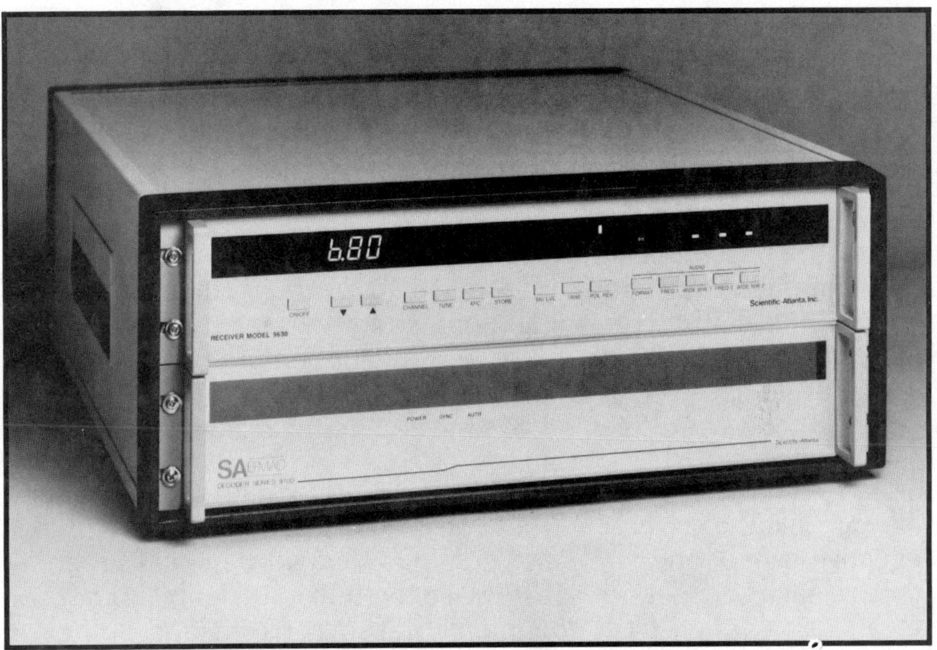

Figure 4-13. Scientific Atlanta B-MAC Receiver and Decoder. *The 9630 satellite receiver and 9700 decoder are rack mounted so there are no exposed interconnecting cables. (Courtesy of Scientific Atlanta)*

Each decoder/receiver may be authorized to receive any channel, program group, individual program or impulse pay-per-view. Audio channel options associated with each channel can also be selected at will. The flexible addressability feature also allows both text information as well as data transmission to be enabled or disabled at will from the headend computer.

D. Encryption Security

Four basic principles can be identified in creating a secure scrambling system. First, a technique that renders the entertainment value of the programming useless is required. This can be accomplished by a wide variety of methods as outlined earlier in this chapter.

Second, the system should not be defeatable by one-time solutions. For example, removing a trap or installing a bandpass filter to eliminate the effect of an interfering carrier can be easily accomplished once and for all. This criteria implies that more secure scrambling methods will have a time-varying component. Therefore, a system which has pseudo-random line inversion is more difficult to crack than one which inverts each line.

Third, the encryption system should be resistant to those observing the scrambled waveform. This implies, for example, that high security systems should use time shifting or should transmit control and other information in digital form. Security would be compromised in a system using random video inversion which relayed a clearly recognizable "flag" in the vertical interval whenever video was inverted.

Fourth, the system should deactivate if information is not continually received from the control center. Addressability is certainly less effective if only one activation command would be necessary to authorize each decoder permanently. Encryption is no doubt more secure in a system which must continually transmit information on how to decode rather than just on when to decode.

No encryption system is foolproof. Talented and dedicated pirates will always be able to break even the most secure code. However, a well-designed system which has no weak link or "back door" can provide security against all but the most sophisticated tampering. A key factor is also the motivation of the pirates. If broadcasts are available at what is perceived a fair price, the desire to beat the system will certainly diminish.

V. CABLE TV SCRAMBLING SYSTEMS

Scrambling systems were originally developed for cable TV networks. Many cable systems employ TV set top converters which typically convert any selected cable channel onto VHF channels 3 or 4. Scrambling devices are usually mated with or built into the cable TV converter.

Some of the commonly encountered scrambling devices in use today are outlined in this chapter. Please note that this overview of scrambling systems is not all inclusive. For example, a number of companies including Electroline and Delta Benco manufacture addressable taps. Similarly, firms other than Oak have used the sine wave scrambling method. But the systems described here are representative of those in use today.

It should also be realized that some of the systems outlined in this and the following chapter such as the Orion, VideoCipher, Z-TAC or Starcom products are registered or trademarked. However, for brevity, the trademark or registered symbols are not included in the text.

Figure 5-1. Scientific Atlanta Headend Equipment. *A full-blown headend featuring channel processors, modulators and control equipment is shown here. The racks are carefully designed so that all wiring is labeled, bundled and adequately shielded and supported. Scrambling computers and encoders are easily incorporated into such a facility either on-site or remotely located. Each terminal connected to the headend control computer monitors and directs the system. (Courtesy of Scientific Atlanta)*

A. SCIENTIFIC ATLANTA ADDRESSABLE TAP

The Scientific Atlanta addressable tap system consists of three main elements: the central control computer; the cable data transmitter; and the field installed, addressable tap.

The heart of this system is an IBM PC equipped with a floppy disk for permanent data storage. The operator enters tiering and pay-per-view or special event programming information about each subscriber for transmission to all addressable taps on the cable network. The computer software is designed for ease of use by being menu-driven so that a novice can be easily taught the control and data entry functions and operations. This language can be personalized to incorporate operator selected commands. The special event feature of the software enables the operator to pre-load addresses for both prepaid and pay-per-view programming so that a single global command can be used to deactivate and then reactivate the taps at the beginning and end of segments of purchased entertainment. The software also permits scheduling of pay-per-view events on top of an existing tier, so that subscribers who do not normally receive a particular service can watch only the special event on that channel. And the whole system can be linked to billing software for ease of operation and accounting.

Security is enhanced by a "refresher" feature. The headend computer periodically transmits an authorization message to all subscribers. If communication between the headend and addressable tap is interrupted or the refresher command is not transmitted, notch filters will automatically be activated and service will be terminated.

Addressing data is frequency modulated onto an 88.1 MHz subcarrier and is transmitted over the cable network. This frequency is just above the upper limit of channel 6 within the FM radio band which extends from 88 to 108 MHz.

Each addressable tap has four tiers of service which are controlled from the headend computer. Each of these tiers can be either a single

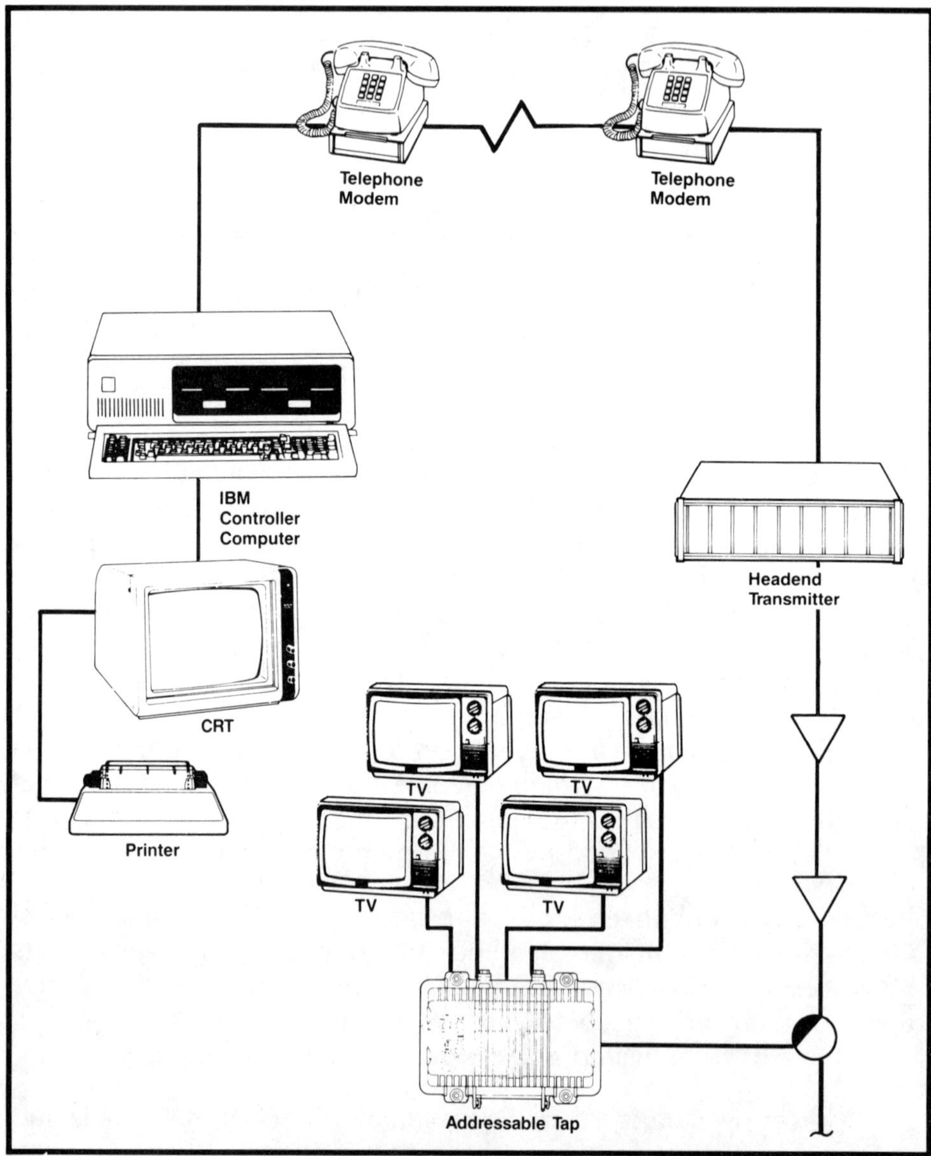

Figure 5-2. Block Diagram of SA Addressable Tap System. *This is a simplified diagram of the Scientific Atlanta Model 2470 addressable tap system. (Courtesy of Scientific Atlanta)*

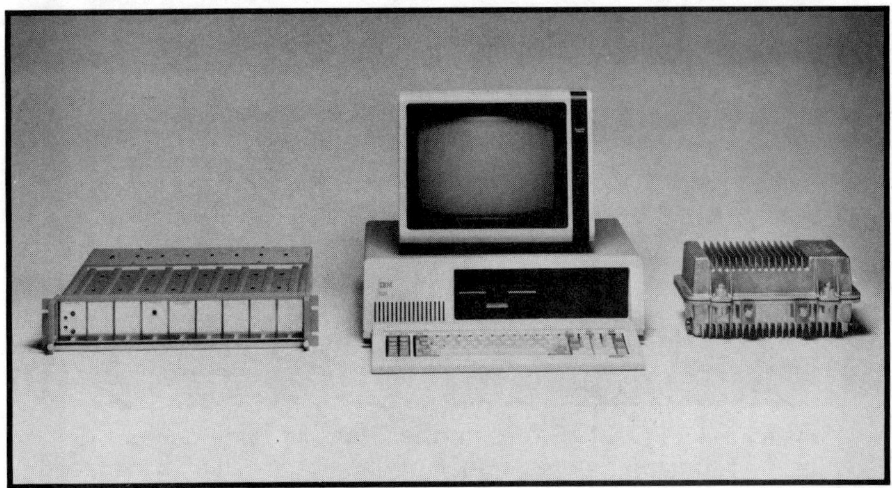

Figure 5-3. Scientific Atlanta Addressable Tap System. *The SA series 2470 addressable tap system is designed to provide low-cost addressability for smaller cable systems offering the operator to change both basic and premium services without visits to subscribers. Television signals are selectively trapped out external to the residence thus improving security. Pictured here from left to right are the cable system transmitter, the IBM PC control unit and one addressable tap. (Courtesy of Scientific Atlanta)*

channel or a range of contiguous channels. The system is designed to manage up to 4,000 taps each having four subscriber ports for a total of 16,000 addresses. Channels 2 through 13 are used to relay all programming so that service is selected by the TV tuner and a cable converter is not required. Such systems are often used in hotels and motels for pay-per-view of movie channels.

B. PICO OTAS ADDRESSABLE TAP

Like the SA addressable tap system, the Pico OTAS (Outdoor Terminal Addressable System) controls subscriber authorization by controlling the video and audio signals in off-premise traps. The philosophy behind this type of system is opposite to that followed in

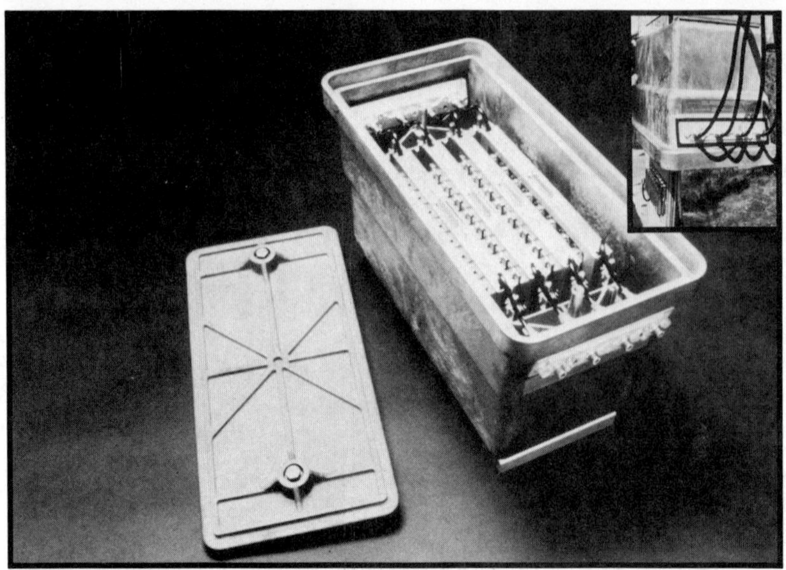

Figure 5-4. Pico OTAS 4-Subscriber Pole-Mounted Cabinet. *This cabinet contains four subscriber modules each of which manages eight tiers of programming. The cabinet not only provides security against theft and tampering because the unit is mounted in an inaccessible out-door location, but also is weather resistant and shielded from RF radiation. Subscriber addresses are actually written onto each module after installation by signals from the control computer which burns in fuses indicating a digital zero or one. (Courtesy of Pico Products, Inc.)*

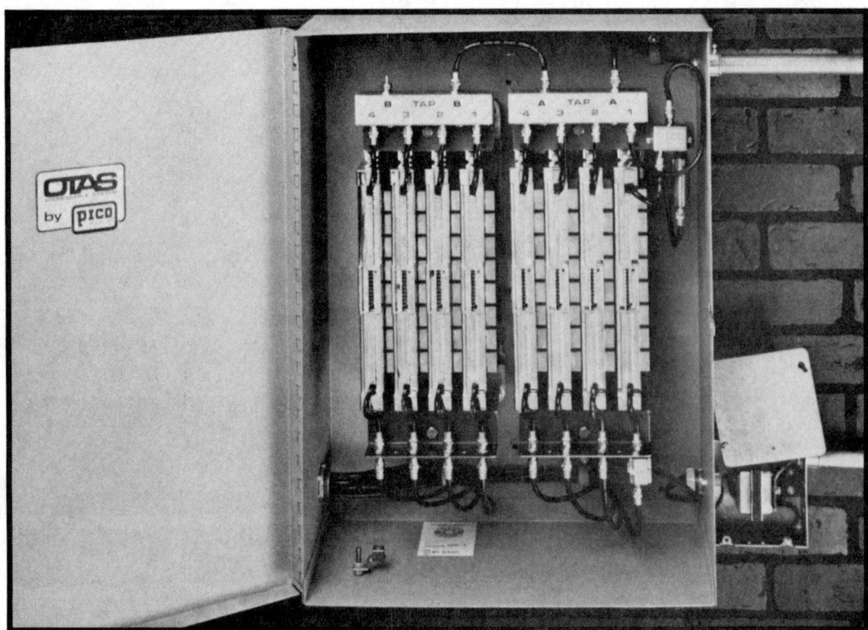

Figure 5-5. OTAS Multiple Dwelling Security Cabinet. *This enclosure accommodates 8 subscriber modules for installation either in utility rooms or outdoors. (Courtesy of Pico Products, Inc.)*

most other types of cable scrambling systems. Instead of creating a signal that needs to be decoded, system logic requires that the encoder outside the customer residence must be bypassed to unscramble the signal.

OTAS equipment consists of up to 8 and 40 subscriber modules installed in security cabinets or racks, respectively. Each module provides up to 8 levels of service including pay-TV programs and the basic fare. The seven tiers of pay-TV plus basic service offered to each customer can be single or multiple channels, including entire bands. With the exception of a plug-in power supply, all hardware is located outside the subscriber's home for security against theft and tampering.

Up to 157,464 subscriber terminals can be addressed per system. The control and addressing data is carried on an out-band, 103.8 MHz carrier at a rate of 10,000 to 16,000 subscribers per minute. Data in the two-way system is returned to the control computer on a 24.3 MHz carrier. The system is designed so that each subscriber terminal must receive its correct address twice in succession in order to be activated. This scheme reduces addressing errors to negligible amounts. Subscriber modules consist of an address/command decoder, a 9 bit data encoder and an 8 bit addressable latch coupled to 8 pin diode switches. These gates are used to select any combination of the eight program tiers. When the correct address is received, the latch opens any combination of the 8 pin diode latches and allows clear television signals to pass. The active trap accomplishes scrambling by strongly overwriting the video signal with two interfering carriers. One is tuned to the video carrier center frequency of the channel; the other varies at slightly less than 3 times the horizontal scan frequency (47 MHz) continuously around the frequency of the fixed interfering carrier. The resulting signal is dramatically altered and, for all purposes, rendered useless at the television receiver. Audio information is also scrambled by remodulating it onto the same 47 MHz interfering carrier. This signal is divided by factors of 32, 64 and 4096 to create audible annoying tones on the television receiver.

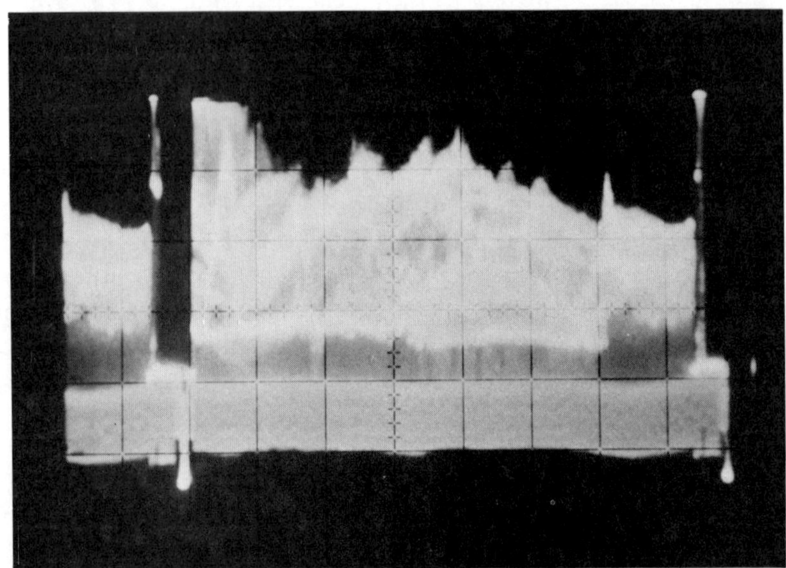

Figure 5-6. Effect of the Pico Active Trap. *The video signal is highly distorted by the Pico active trap. Two interfering carriers, one fixed and one varying, accomplish this distortion. (Courtesy of Pico Products, Inc.)*

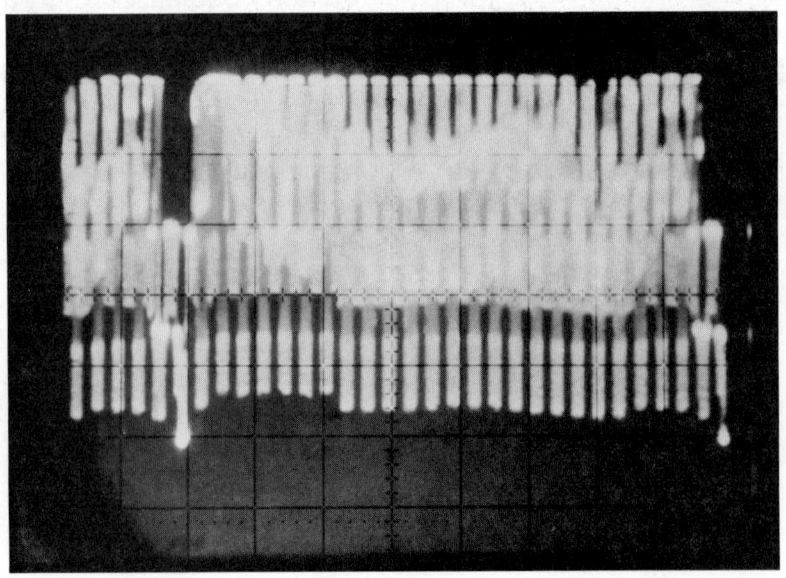

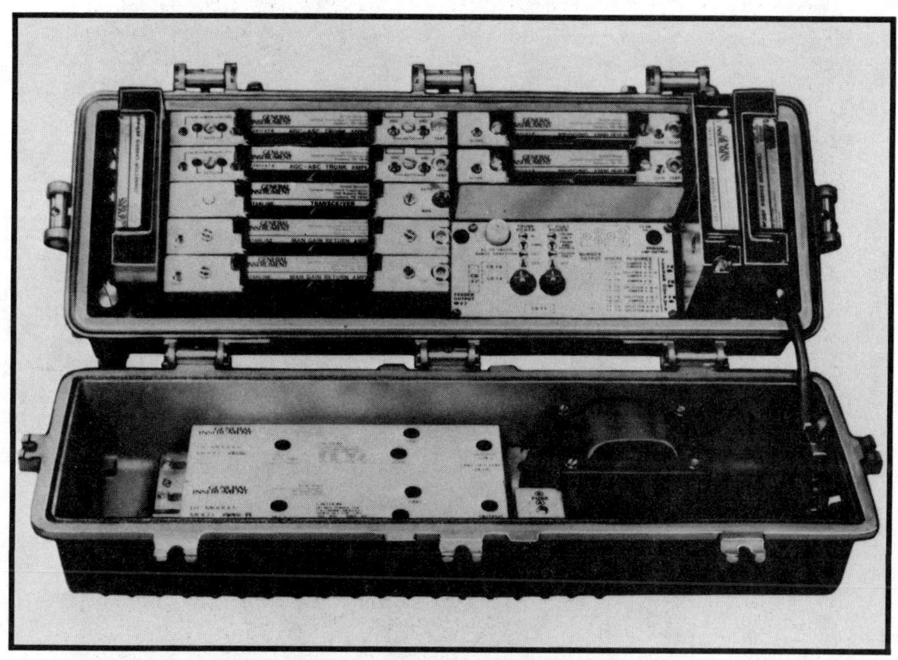

Figure 5-7. Jerrold Starline 400 Line Amplifier. *Both scrambled and "clear" signals pass equally well through this trunk/bridger amplifier. This unit is capable of feeding status information back to the cable headend from decoder boxes in a two-way system. (Courtesy of General Instrument)*

Figure 5-8. Jerrold FFT-Series Tap. *This unit is a passive device, in contrast to an addressable tap. It permits signals from a cable distribution line to be "tapped" into one or more residences. It passes frequencies in the 5 to 400 MHz range with a specified insertion loss and port output. Taps are often installed outdoors for extra protection from tampering. This variety has a weatherproof aluminum enclosure with molded neoprene gaskets and self-sealing "F" ports to prevent entry of moisture and dirt. (Courtesy of General Instrument)*

C. GATED SUPPRESSED SYNC

When the gated sync suppressed scrambling is used, the entire horizontal blanking interval and sync pulse in the video signal are suppressed by approximately 6 decibels. This causes the scan lines to begin at somewhat arbitrary positions and results in a confusing series of wavy lines on the screen. The term gated refers to the method by which the suppression occurs. Pulses are added to the composite video signal only during the horizontal blanking period or "gated" in order to lower the voltage level.

There are three common forms of gated sync suppressed scrambling: the in-band; the out-band and the Zenith SSAVI system. The in-band is commonly referred to as the Jerrold and Sylvania computerized addressable system while the out-band is called the Hamlin system. All forms of the SSAVI encryption, today usually marketed under the brand name Z-TAC, are categorized as the Zenith system.

In-Band System

The in-band system contains the reference signal necessary to reconstruct both picture and sound within the standard 6 MHz television channel bandwidth. Pulses in the vertical blanking interval are not altered so that the picture retains vertical stability and does not roll. Although the audio signal is rarely scrambled, when this occurs it can be found modulated onto a 32 KHz subcarrier at the FM sound detector output in the decoder. (This technique is generally only encountered on scrambled UHF broadcasts, known as STV or subscription television.)

Reconstruction of horizontal blanking pulses is straightforward. The reference signal required is a 15.734 KHz pilot tone, equal to the color line scan rate. It is frequency modulated onto the audio carrier which is always located at 4.5 MHz above the video center frequency on every NTSC television channel. This pilot tone is present in addition to the 32 KHz audio subcarrier. After the required pulse is reconstructed, it is gated via an amplifier or mixed with the baseband video to increase the voltage of the supressed blanking and sync pulse.

Out-Band System

The reference signal in an out-band system is amplitude modulated onto a carrier on another television channel outside the active video channel. Reconstruction follows the same methods used in in-band sync suppressed decoders.

The location of the pilot tone can be determined by finding a channel which displays a clean white raster. Note that this pilot or reference signal is never included on UHF channels 30 through 34. The pilot center frequency is typically found on UHF channels 42, 49, 50, 52 and 53.

D. OAK SINEWAVE

Sinewave scrambling systems, a typical example being the Oak Sinewave, alter the video signal by resistively summing a sine wave that is locked in both phase and frequency with the composite baseband video. As a result of the ensuing distortion, the television receiver circuit which normally detects and separates the sync pulses, known as the sync separator, cannot recognize the modified signal. This separator as well as the automatic frequency control (AFC) circuit mistakenly attempts to lock onto the peak of the sine wave instead of the horizontal sync pulse. The result is a wavy black and white line surrounded by color distorted video which appears down the center of the television screen.

The sine wave is suppressed during the vertical blanking interval so that it has no effect. Therefore, the picture retains vertical stability and does not roll.

Decoding the video is rather simple. The sine wave is detected, shifted in phase 180 degrees and summed back into the composite baseband video signal. Adding this second out-of-phase sine wave perfectly cancels the first one and restores the signal to its original form. The restoring sine wave must be suppressed like the original scrambling sine wave was during the vertical blanking interval.

Some of the Oak Sinewave type encryption systems use the variable sync (VS) modification. These VS enhanced scramblers use a sine wave that is double the normal line scanning 15.734 KHz rate or alternate back and forth between the normal and higher 31.468 MHz frequency rate.

The Oak Sinewave system gives a broadcaster the option to scramble the audio as well as the video information. This is accomplished by frequency modulating a 62.5 KHz subcarrier with the audio signal. When scrambling is not used, the audio signal directly modulates the normal audio subcarrier. In contrast, the scrambler uses this pre-modulated 62.5 KHz subcarrier instead of the raw audio to modulate the normal subcarrier. This signal is therefore frequency modulated twice. Television receiver circuitry does not recapture the raw audio but a 62.5 KHz FM signal which will not produce a recognizable sound on the television speakers. In order to decode the audio a second FM demodulator present in the descrambler is needed to extract the audio and convert it back to the proper subcarrier location, 4.5 MHz above the video center frequency.

E. ZENITH SSAVI

SSAVI (synchronization suppression and active video inversion), was developed by the American Television and Communications Corporation (ATC) one of the largest cable TV companies in the United States, and the Zenith Electronics Corporation. It was for years the most complex scrambling system in use for both cable TV and over-the-air broadcasting. This technology has also been featured as a lower-cost option on some satellite TV networks. Decoders are sold under the brand name Z-TAC for both cable and over-the-air broadcasts.

Video Scrambling

The SSAVI system offers 6 levels of security. In addition to suppressing the horizontal sync pulse, the encoder uses video inversion on a "pseudo-random" basis to improve overall signal security. While the switching rate between inversion modes can be truly random and as fast as every frame, it is usually changed on a daily to monthly or on a pseudo-random basis in response to variations in picture structure. This facilitates the defeat of pirate decoders that are not capable of dynamic mode switching. It is interesting to realize that when "adult" programming is transmitted the SSAVI usually uses the fastest normal/inverted switching mode available.

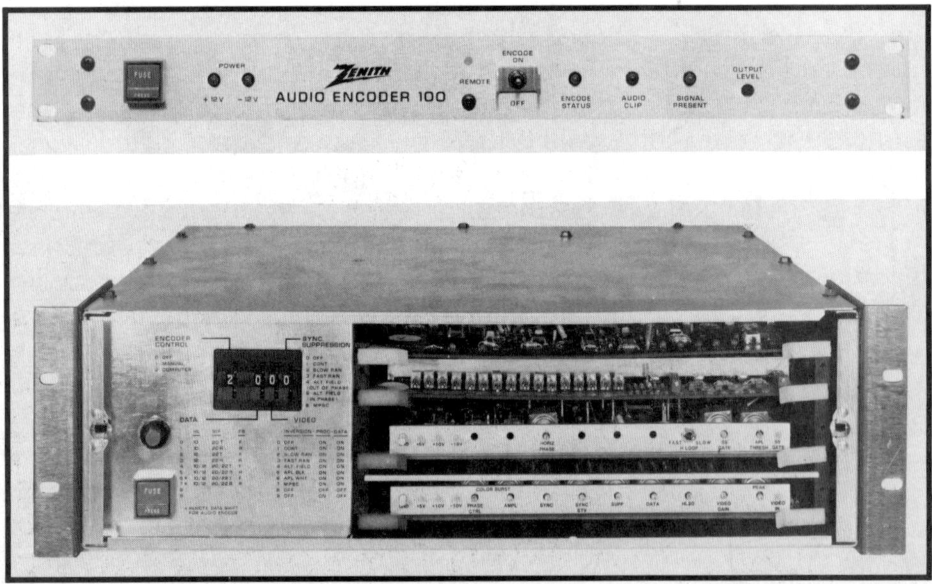

Figure 5-9. Zenith SSAVI Encoders. *The ZVE-110 video and ZAE-1000 audio encoders are rack mounted and fed into the modulator IF loop at the cable headend. These encode outgoing television signals in addition to adding address and program tag information onto the outgoing signal. (Courtesy of Zenith Electronics Corporation)*

The six levels of security which can be chosen at will by the operator are permutations of sync suppression on and off, video inversion on a pseudo-random basis, and video inversion as a function of average picture level (APL) changes. The encoder may be set to one of two APL modes, black or white. APL black causes only bright scenes to be inverted. Since all transmitted fields are dark, the video waveform is maintained well above the suppressed sync pulses. Increased tearing of the picture is the result. Similarly, APL white causes only dark scenes to be inverted. Since bright pictures correspond to lower RF voltage levels, use of APL white results in a welcome reduction in required average signal power. The six levels of security are:

(1) Suppressed sync and inverted video (pseudo-random basis)

(2) Suppressed sync and inverted video (APL basis)

(3) Suppressed sync and normal video

(4) Normal sync and inverted video (pseudo-random basis)

(5) Normal sync and inverted video (APL basis)

(6) Normal sync and normal video

No in-band or out-band reference signal is transmitted to facilitate reconstruction of the sync pulses in order to enhance scrambling security. An oscillator running at the line scan frequency of 15.734 KHz, which is synchronized with the vertical sync pulses and free running during the rest of the picture field, serves as a reference timer for restoration. However, not all the horizontal pulses are suppressed. The scrambled signal still contains horizontal sync pulses from about 2 lines prior to vertical blanking to line 26 of the following field.

When the inverted mode is selected only the video signal portion is flipped. The end result is that the color sync burst is 180 degrees out-of-phase with the color information in the video signal. Therefore, during the decoding operation only the picture segment of the signal must be inverted.

To restore the video, if one of the inversion modes has been selected, it is necessary to have the video signal available in both the normal and inverted form with a common DC point. This circuit detail allows use of a pair of relay gates so that the appropriate video signal can be selected. Encrypted data in the vertical blanking interval signals the decoder whether to select the normal or inverted video signal path.

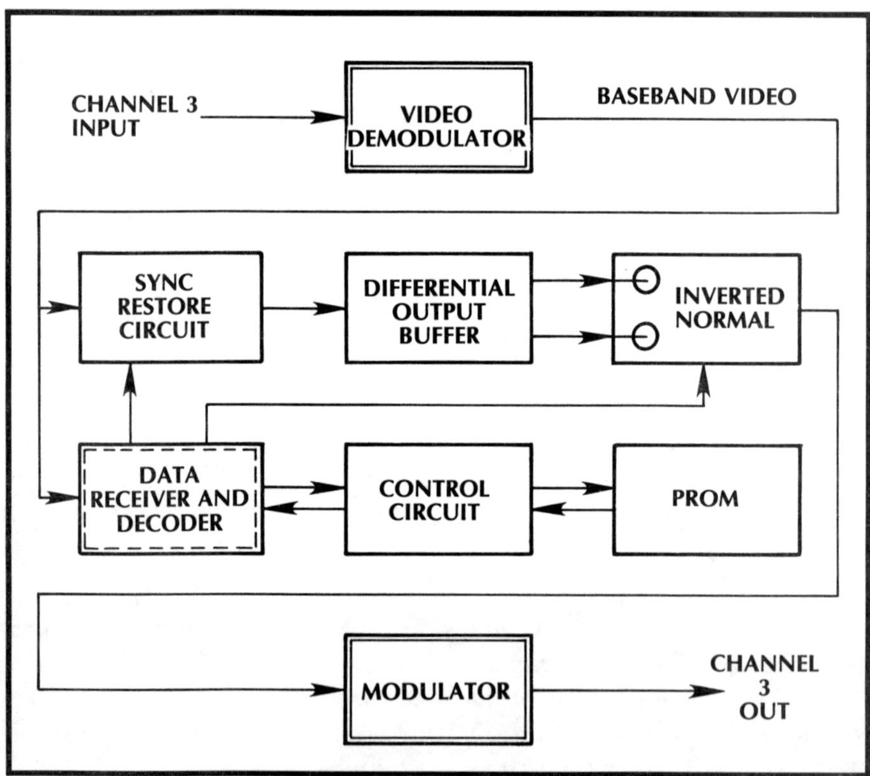

Figure 5-10. SSAVI Baseband Decoder - Block Diagram. *This simplified diagram shows the logic behind decoding a SSAVI encoder signal. An algorithm contained in a programmable read only memory (PROM) reads the data received within the baseband audio signal and directs a sync restore circuit. The data receiver and decoder controls a gate which selects between inverted and normal video for relay to a channel 3 modulator.*

Audio Scrambling

The audio portion of either cable or over-the-air broadcasts may be scrambled by a "spectrum inversion" technique. The baseband audio spectrum is reversed so that a 1 KHz signal is transposed to become approximately 14 KHz. In earlier versions, over-the-air and satellite broadcasts incorporate the same method used by the Oak Sinewave encoding of audio with the difference being that the FM modulated subcarrier is set at 40 KHz rather the 62 KHz.

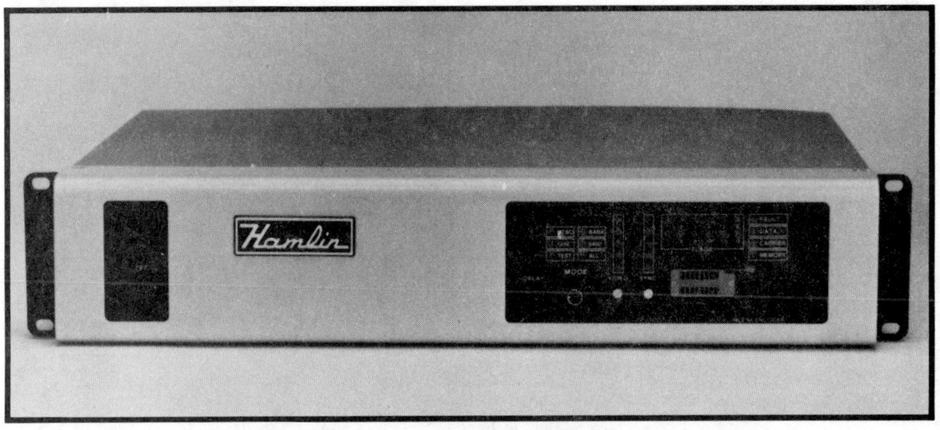

Figure 5-11. Hamlin MLE-64 Multi-Level Encoder. *The Hamlin scrambling system is similar to the SAAVI because sync suppression is used for video scrambling. The horizontal sync is reconstructed by locking onto the equalization pulses in the vertical blanking interval. This unit has front panel displays of video and sync levels and is upgradable to the addressable Model MLE-64A.*

Addressing is accomplished by inserting 5 lines of 24 bits each into the vertical blanking interval following the equalization pulses. Bits can be used in the following ways. Sync data is divided into an 8 bit command word describing the message function and 8 bits for channel identification. Code manipulation necessary for descrambling is allocated 8 bits. A 16 bit word designates the 65,536 groups of customers, each of which contains 144 customers. If the customer addresses matches the particular terminal, the command word is acted upon. (Courtesy of Anixter Communications)

Addressability

SSAVI decoders are fully addressable allowing the operator to authorize or terminate service without the need for a site visit. Addressing information is contained in a PROM powered by a rechargeable battery so that memory is retained during power outages or times when the decoder is unplugged.

Figure 5-12. Zenith Z-TAC Controller. *This Hewlett Packard HP-1000 computer is used to manage the Z-TAC pay-per-view (PPV) cable TV scrambling system. From top to bottom the rack contains a SSAVI ZVE-110 encoder, a Z-VIEW return data receiver and an HP computer. This system is an upgrade to the original Phonevision which manages a PPV system which accepts feedback over telephone lines. The Z-VIEW system is a true two-way impulse PPV system whereby customers can choose programming by pushing the appropriate buttons on their remote controlled decoder. The telephone data receiver used in the Phonevision system is here replaced with a return data receiver and an additional module, shown in front of the rack, to be added to the Z-TAC decoder. (Courtesy of Zenith Electronics Corporation)*

There are two codes used to identify the individual channel, the market code and program tag, and one used to change customer service levels, the address. When broadcasts are scrambled, the unit will function only if its pre-established market code matches the one being transmitted by the cable operator. A unique market code is assigned to each cable system so that swapping unauthorized decoders between different cable networks is inhibited. SSAVI decoders are addressed with program tags corresponding to purchased scrambled services. At the time of installation, each encoded channel is given one of the program tags via the controller. Like the market code, if the incoming program tag is properly matched to the one stored in memory the unit will function and decipher the channel in question. Customers have access to a total of 20 program tags or tiers. These are organized into five categories in each of four channel banks.

Each decoder also contains a unique serial number so that the headend computer can identify and address any customer at will. Whenever cable operators detect an unregistered serial number the decoder is deactivated. Only an unmodulated video carrier is received and the television screen goes black and sound disappears.

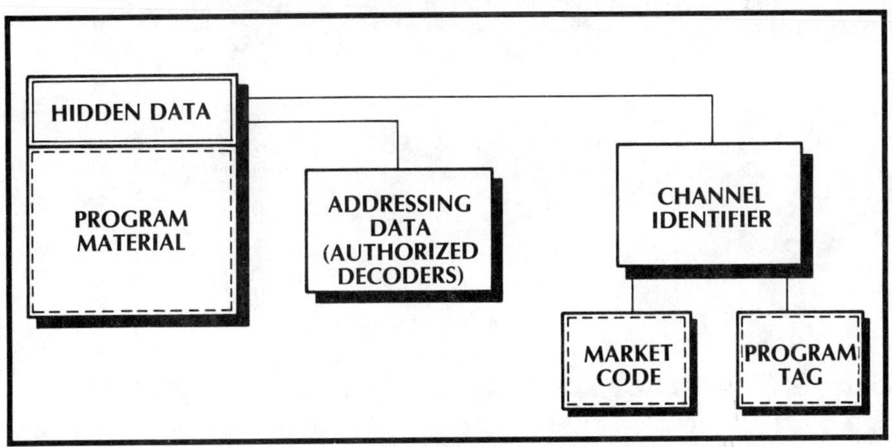

Figure 5-13. Z-TAC Addressing Format. *In-band data transmitted from a Z-TAC headend contains addressing data for decoder identification and authorization and a channel identifier. The latter digital stream includes both the market code and program tag.*

Addressing and command information for this encryption system is broadcasted on lines 10, 11 and 12 in the vertical blanking interval of each field. Program and market code commands are transmitted on line 13.

There are four components at the heart of a Z-TAC system, the controller, the encoder, the modulator and the decoder. Software writ-

Figure 5-14. Zenith 1600 Series Baseband Addressable Decoders. *This remote controlled decoder series is designed to be used with the Z-TAC scrambling system. Many features are available including parental control/favorite channel memory, BTSC stereo compatibility, flashback and favorite channel scan. An external connection allows for teletext and two-way communication expansion. (Courtesy of Zenith Electronics Corporation).*

ten for the four available computers has the capability to manage four tasks: configure the system to the operator's needs; assign program tags to the encoders; assign controller access passwords; and manipulate the global list of subscribers. When the controller is idle, it automatically transmits the entire list of subscribers and their permissible tags. The computer is designed so that it can be linked to an external billing and management system.

The encoder accomplishes two tasks. It adds the addressing and program authorization data to the video signal. And then it manages the audio and video scrambling. The modulator processes the composite baseband signal onto the proper RF channel frequency. The decoder is integrated into a standard cable TV-type converter.

A text module can be employed in conjunction with the Z-TAC decoder to allow delivery of a one-way teletext service. Either of two pay-per-view two-way systems may be used. One employs an upstream transmitter operating at 5.5 and 11 MHz. A small add-on module is connected to an interface jack on the rear of the decoder. The other is a high speed telephone based system requiring no additional equipment in the home.

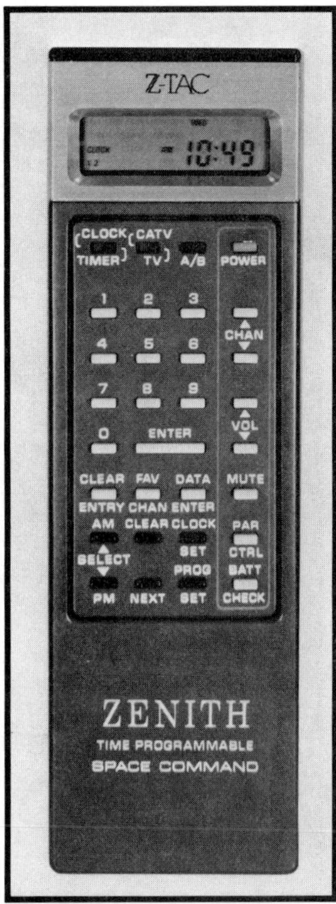

Figure 5-15. Zenith TAC-Timer. *This infrared remote control can be programmed to change channels on the Z-TAC decoder at specified times so that a VCR will automatically record when programmed for time and channel. (Courtesy of Zenith Electronics Corporation)*

Figure 5-16. Result of Z-TAC Scrambling. *This photo shows what appears on screen as a result of Z-TAC scrambling. (Courtesy of Zenith Electronics Corporation).*

F. TOCOM

The Tocom system is quite similar to the Zenith Z-TAC in providing baseband addressable scrambling. Three levels of video protection are available: partial video inversion and compression; random dynamic sync deletion; and random dynamic video inversion. Data to recover the scrambled video is encrypted and inserted onto lines 17 and 18 of the vertical blanking interval. This 92-bit data packet is transmitted just like the picture portion of the video signal over a 53 microsecond interval following the color burst. The audio signal is sent in the clear.

The headend is designed around an IBM AT personal computer which can alone manage up to 32,000 subscribers. However, the capacity of the entire system can be expanded to over 250,000. Software is supplied with a built-in database containing a structure to manage demographic information about each subscriber. The database can be easily manipulated from up to three independent terminals. A separate user-owned billing computer can feed the IBM "slave."

Figure 5-17. Tocom Plus Micro-ACS 5521 Headend Computer. *This headend computer is intended for control of smaller systems where the number of subscribers is less than 32,000. This unit feeds the Tocom encoder. (Courtesy of General Instrument).*

The baseband addressable converter has a number of attractive functions. Up to 66 cable channels can be selected and 32 tiers of scrambled programs are controlled from the headend. A built-in timer allows up to four events programmed over a week period to be fed via composite audio/video baseband outputs to a VCR. The choice between a television or VCR output is controlled by an internal electronic A/B switch. The addressable converters have other features including parental lockout, diagnostic messages, on-screen text and graphics capability, channel number and time recall and preferred channel selection. Teletext can be relayed either on lines 17 and 18 of the vertical blanking interval, sharing those lines with addressing and program control information, or as full-channel text.

Figure 5-18. Tocom Plus Headend Video Processor. *This unit scrambles up to eight cable channels per rack-mounted enclosure. Coupled with the headend computer it dynamically controls addressability and access to scrambled channels. Converters can be addressed even when no video signals are present such as after sign-off time because an internal video generator is provided. (Courtesy of General Instrument).*

Figure 5-19. Tocom Plus 5503-VR Converter. *This remote controlled converter features 7-day, 4-event programming, parental lockout, preferred channel memory with last channel recall, self diagnostics, 66 video channel selection and a 32 channel tier structure. Authorization is updated every few seconds by refresher signals to prevent a timed disable process. This converter, one component in the Tocom 55 Plus series, can be upgraded by adding extra modules to receive BTSC stereo or teletext. (Courtesy of General Instrument).*

When the interactive version of this system is chosen, it is structured to allow consumer control of impulse pay-per-view (IPPV). Once a preauthorized credit limit has been established, a subscriber can view a pay-per-view event by simply entering a personal code via the remote control. At any time thereafter, the headend computer polls and instructs all converters to relay the credit register information via a telephone digital dialer so that correct billing can be established.

G. JERROLD STARCOM

The Jerrold Starcom addressable system is designed to provide RF sync suppression scrambling at a moderate level of security. There are three possible video scrambling states: 6 dB sync suppression; 10 dB sync suppression; and 6 or 10 dB sync suppression. The highest security level is provided by the variable suppression level scrambling which can be manually selected by the operator at each of the preset switching intervals. This pseudo-random method can be broken by use of a sync regeneration circuit.

One of two headend computers manage the addressable system. The one-way AH-2E controller is built around a DEC Micro/PDP-11 computer. Two types of addressing information are transmitted to each customer's converter/descrambler: an identification code defines program tiers and is encoded onto each scrambled channel; an authorization address unique to each decoder is relayed on an out-band RF carrier. The AH-2E supports 32,000 decoders but has an upper limit of 224,000 subscribers.

The AH-4E controller, structured around a DEC PDP-11/23 PLUS minicomputer, is designed to manage a two-way addressable system via as many as eight terminals. In the basic configuration 32,000 subscribers can be controlled; up to 256,000 can be managed in the fully expanded state. It establishes two-way communications with the Starcom V two-way terminal via the encoder, thus accomodating services such as impulse pay-per-view (IPPV) and opinion polling. When used with the one-way Starcom 450, its return communications are limited to phone lines via add-on modules. Downstream communications are transmitted on a 106.5 MHz carrier; upstream data is relayed by a carrier in the 8.3 to 10.4 MHz band below VHF channel 2.

The Starcom 450 and V converters are used with one-way and two-way systems, respectively. A series of add-ons includes upgrades to stereo, or IPPV via phone lines.

Figure 5-20. Jerrold Starcom DS/E-1000 Digital Scrambler/Encoder. *This unit scrambles the RF headend signal by suppressing the horizontal sync pulse. Signal processing occurs as follows. The baseband composite video enters the DS/E-1000 for sync detection and is then routed to the modulator IF loop. The modulator provides video and audio IF signals to the encoder. The sync suppressed video and clear audio is returned for modulation onto any chosen television channel. (Courtesy of General Instrument)*

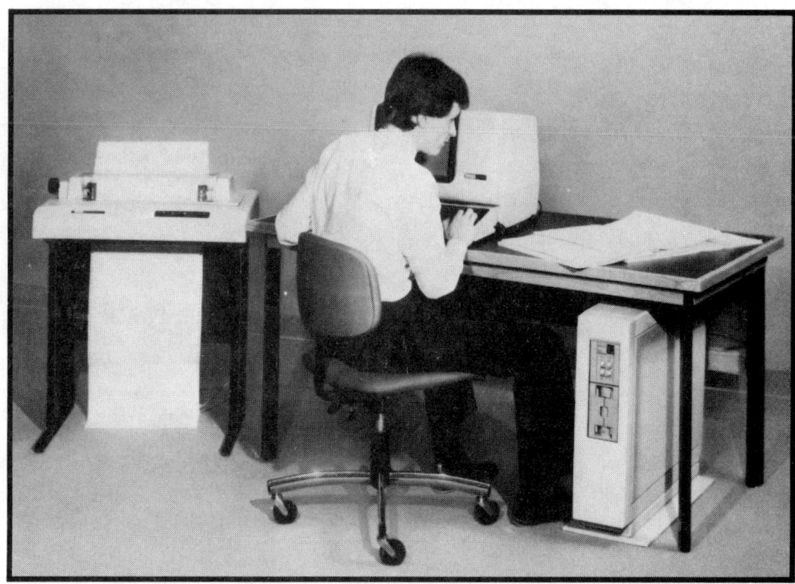

Figure 5-21. Jerrold Starcom AH-2E Addressable Control Center. *This headend control center includes the computer and its associated peripherals, software package, addressable encoders and the necessary supporting equipment such as a printer for hookup to a second computer for billing and other services. (Courtesy of General Instrument)*

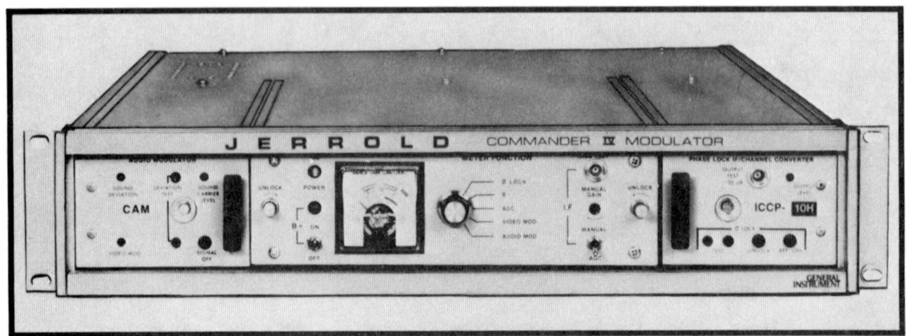

Figure 5-22. Jerrold Commander IV Modulator. *The Jerrold Starcom encoder can be matched with this modulator in order to accomplish the necessary RF sync suppression scrambling. (Courtesy of General Instrument)*

Figure 5-23. Jerrold Starcom VI Addressable Cable Converter. *This decoder is an upgrade of the Starcom 450 RF addressable converter for use in one-way scrambling systems. It is capable of 66 channel reception and can be expanded to twice this number. It features all the standard customer features as well as parental lockout, last channel recall, favorite channel programming, built-in diagnostics and remote control capability. (Courtesy of General Instrument)*

Figure 5-24. Jerrold Add-on Starfone IPPV Transmitter. *The Starfone impulse pay-per-view transmitter upgrades the one-way Starcom 450 or VI addressable converter into one which manages return communications via conventional telephone lines. (Courtesy of General Instrument)*

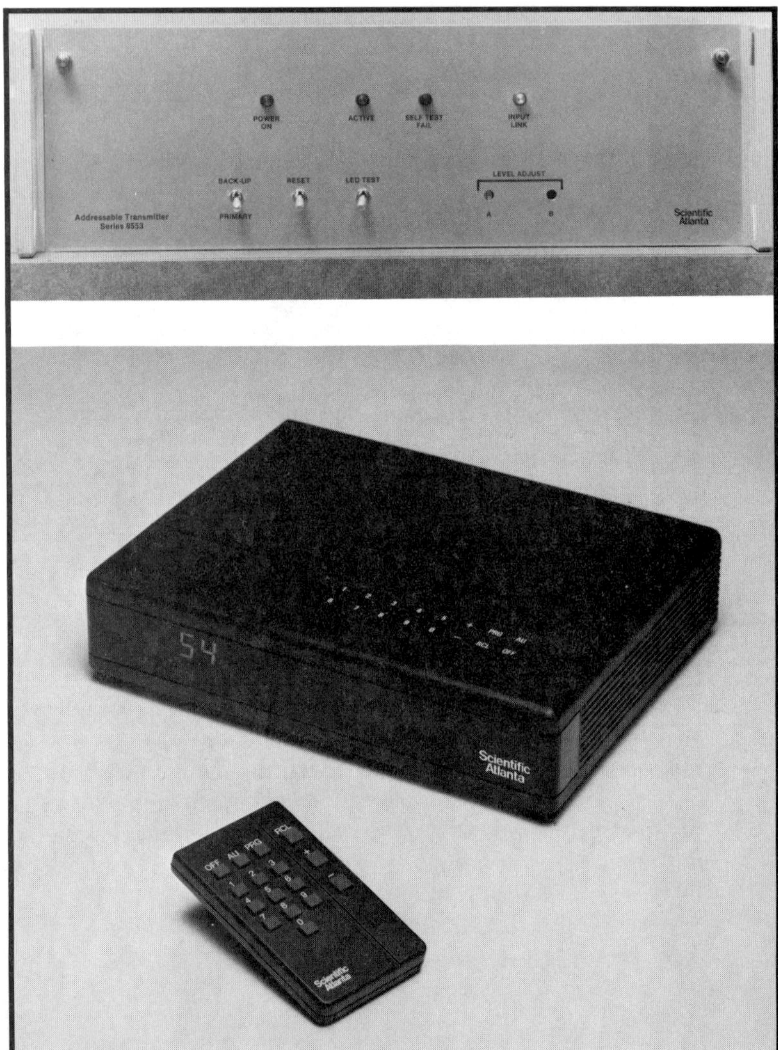

Figure 5-25. The Scientific Atlanta Series 8500 Addressable Transmitter and Decoder. *These components are members of the SA 8500 scrambling system which, like the Jerrold Starcom system, uses 6 dB dynamic suppressed sync video. It is expandable to a fully addressable system. The addressable transmitter operates at a standard frequency of 108.2 KHz, just at the upper end of the FM band, and is fed by the control computer. It can address individual or groups of decoders at once and must send a refresher signal to reactivate decoders at time intervals ranging from half an hour to 64 hours. (Courtesy of Scientific Atlanta)*

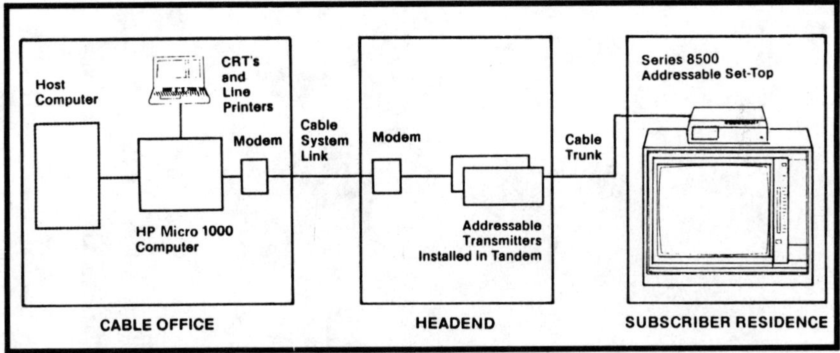

This is a simplified drawing of a scrambling system with the control computer located in another building removed from the headend. Computer and transmitters are interfaced through telephone modems. (Courtesy of Scientific Atlanta)

Figure 5-26a. SA Addressable System Manager and Control Unit. *The System Manager, used with the Scientific Atlanta 8500 series of scrambling products, handles all addressing and control functions including pay-per-view. The central component is an HP Micro 1000 computer. Individual units can be addressed to begin service, change service levels, authorize single channels or tiers, disable any unit or relay global addressing commands. The System Manager feeds information into the Addressable Control Unit and then via a modem to the cable network. Multiple headends can be served each requiring one addressable transmitter. (Courtesy of Scientific Atlanta)*

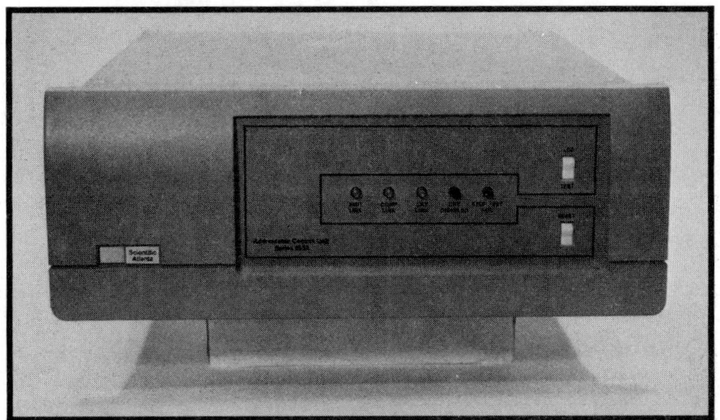

Figure 5-26b. *See Figure 5-26a caption.*

H. OAK SIGMA

The Oak Sigma is one of the highest security, cable TV encryption systems available today. In fact, the similarity between the Sigma and the sophisticated Oak Orion satellite scrambling system is apparent by comparing features of both (see Chapter VI).

Three levels of signal protection at baseband are created. The audio and video are scrambled and the descrambling keys are encrypted. The video signal is randomly inverted on the basis of scene change monitoring. This avoids unnecessary flicker and maintains the quality of the descrambled picture. Both the vertical and horizontal vertical blanking intervals are also eliminated. Two digitized channels of audio are encrypted and inserted in the horizontal blanking interval for transmission. The standard audio subcarrier is not used, but is available if necessary.

Two separate control channels are used. A global channel is FSK (frequency shift key) modulated on either of two subcarriers at 104.75 and 112.7 MHz in the FM band. It carries the keys required by the decoder to restore the original signal, the general authorization information, as well as system-oriented control data. It is continuously relayed as a 64-bit structure to both the home terminal unit and the encoders. The second control channel located in the vertical blanking interval contains specific program information relevant to a given television channel and time.

The system is fully addressable. Each decoder has a set of keys, permanently stored in an EEPROM (electrically erasable programmable read-only memory). A decoder is authorized by a matching key being transmitted from the cable headend. Unless this authorization is "refreshed," the decoder will deauthorize automatically after a specified period of time. This addressability feature offers cable operators the ability to control "box swapping" whereby decoders are illegally transferred between customers.

A multi-level key distribution system based on three categories of keys is employed. Each decoder has a secret and unique "box- specific" key, a variable second-level "market" key which is common to all authorized subscribers and service keys. These service keys control the tiering structure and are used to unlock specific channels. These are varied continuously. Each decoder also has a non-secret box address used by the headend computer for routine communications.

There are two layers of tiering control. Each subscriber's decoder can be enabled to receive up to 28 independent program tiers. The tiering structure can be communicated to one decoder. Alternatively, keys relating to groups of decoders can be embedded into code and controlled by a single command. Software is available so that the entire system can be managed by either an IBM or DEC series computer. This command language is designed to interface with the cable operator's existing business computer so that the transition to the Sigma system can be accomplished as smoothly as possible.

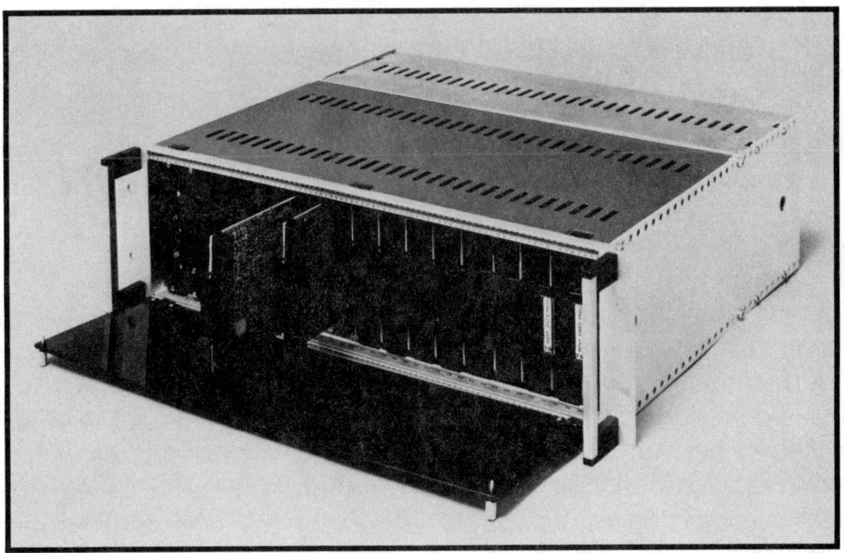

Figure 5-27. Oak Sigma Encoder. *This encoder provides digital audio encryption and sync removal/video inversion scrambling. It also accept data for insertion into the vertical blanking interval. (Courtesy of Oak Industries)*

Figure 5-28. Sigma One Home Terminal. *This remote controlled decoder processes the highly encrypted Oak Sigma signal. It has a wide range of features including parental lockout, last channel recall and remote on/off. (Courtesy of Oak Industries)*

I. MULTIPOINT DISTRIBUTION SERVICES

Multipoint distribution services (MDS) are analogous to over-the-air cable TV networks. Satellite broadcasts which are received at a central facility are retransmitted within the 2150 to 2162 and 2500 to 2680 MHz frequency bands to small roof-top antennas in the vicinity. Each subscriber antenna must have a clear line-of-sight view to the transmitting antenna because trees, buildings or other obstructions strongly absorb the microwaves. The range of a MDS transmitting antenna varies from 30 miles under average conditions to 90 miles under exception conditions. The rule of thumb is that each watt of broadcast power equals roughly 1 mile of usable reception. MDS transmission power is in the 10 to 100 watt range which is much below the hundreds of kilowatts typically required by a conventional television station.

Until recently, no efforts were made to encrypt MDS broadcasts. However, companies such as General Electric, M/A COM and Zenith have designed and now manufacture decoders specifically for this market. These decoders are basically the same as their cable or satellite TV counterparts. For example, the VideoCipher IV MMDS converter is virtually identical to the VideoCipher II satellite TV decoder except it is packaged differently. The Zenith MMDS security system uses the Zenith Z-TAC decoder technology. These systems are explored in detail in this and Chapter VI. Note that both clear and encoded MDS broadcasts require special detection equipment before being fed into decoders or television sets.

The legality of receiving MDS signals also bridges those issues surrounding satellite and cable TV. Cable TV signals are definitely the property of the parent company and unauthorized reception is illegal. In contrast, broadcasts relayed over-the-air by conventional television stations as well as by satellite can be legally received free-of-charge. However, if satellite programmers decide to encode their transmissions, viewers must subscribe to and pay for their service. MDS operators are regarded not as a broadcasting service but as common carriers, like AT&T, Sprint or MCI. Unauthorized reception of their transmissions is illegal whether or not encryption techniques are used. It is certainly more difficult hiding an unauthorized, roof-top antenna than having an illegal decoder indoors. Operators have successfully prosecuted manufacturers and users of pirate equipment. Viewers intercepting MDS signals can be sued and charged under local anti-piracy laws.

MDS Users and Uses

Multipoint distribution service operators use the authorized microwave frequencies for data communications as well as for television broadcasts. The primary use of the latter is for relaying movies and other pay-per-view programs to hotels and apartment complexes. In hotels guests simply call the main desk to view a movie or special event. Apartment or condo dwellers typically pay $20 per month to view MDS television. While the original systems were capable of transmitting only one channel, new MMDS (Multichannel Multipoint Distribution Service), OFS (Operational Fixed Service) and ITFS (Instruction Television Fixed Service) systems are now available which can manage from 4 up to 31 channels.

TABLE 5-1. FREQUENCY ALLOCATION FOR MMDS AND ITFS SERVICE

ITFS - Groups A, B, C, D and G
MMDS - Groups E and F
OFS - Group H

GROUP	CHANNEL NUMBER	FREQUENCY BAND (MHz)
MDS*	1	2150 - 2156
	2	2156 - 2162
	2A	2156 - 2160
Group A	A1	2500 - 2506
	A2	2512 - 2518
	A3	2524 - 2530
	A4	2536 - 2542
Group B	B1	2506 - 2512
	B2	2518 - 2524
	B3	2530 - 2536
	B4	2542 - 2548
Group C	C1	2548 - 2554
	C2	2560 - 2566
	C3	2572 - 2578
	C4	2584 - 2590
Group D	D1	2554 - 2560
	D2	2566 - 2572
	D3	2578 - 2584
	D4	2590 - 2596
Group E	E1	2596 - 2602
	E2	2608 - 2614
	E3	2620 - 2626
	E4	2632 - 2638
Group F	F1	2602 - 2608
	F2	2614 - 2620
	F3	2626 - 2632
	F4	2638 - 2644
Group G	G1	2644 - 2650
	G2	2656 - 2662
	G3	2668 - 2674
	G4	2680 - 2686
Group H	H1	2650 - 2656
	H2	2662 - 2666
	H3	2674 - 2680
	H4	not assigned

*Note audio and video carriers are inverted

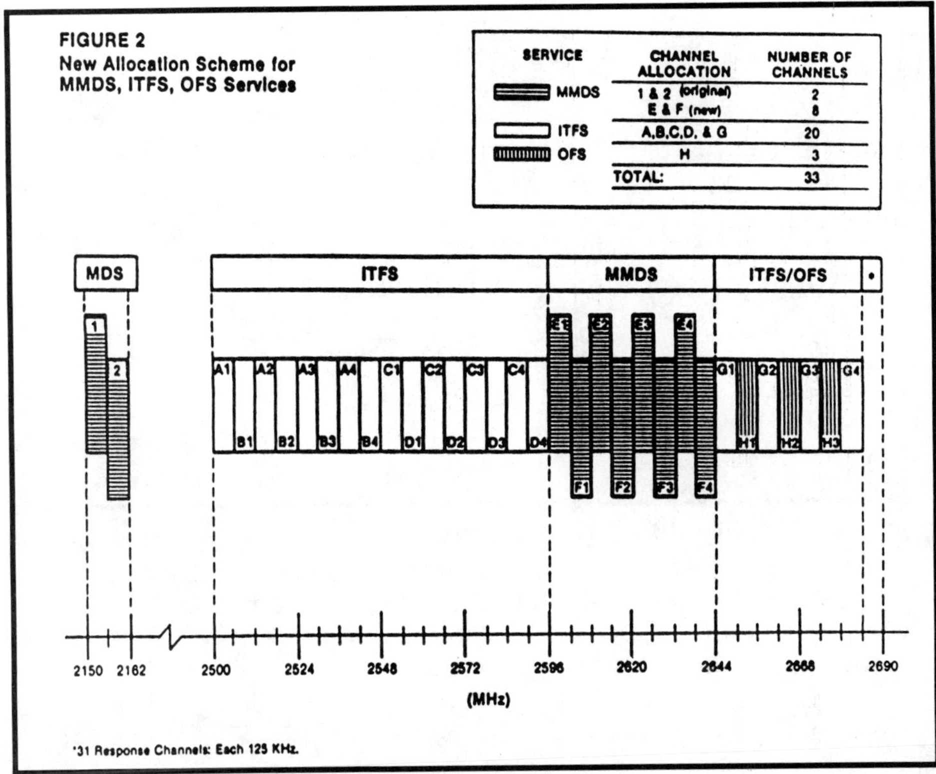

Figure 5-29. MMDS/ITFS/OFS Channel Allocations. *(Courtesy of EMCEE Broadcast Products)*

System Operation

Antennas to receive MDS transmissions can be either parabolic or rod shaped. The latter is like a piece of sewer pipe with a funnel at its top. Parabolic antennas or "dishes" are typically about two feet in diameter and usually provide higher quality reception than the rod type because they have higher gain.

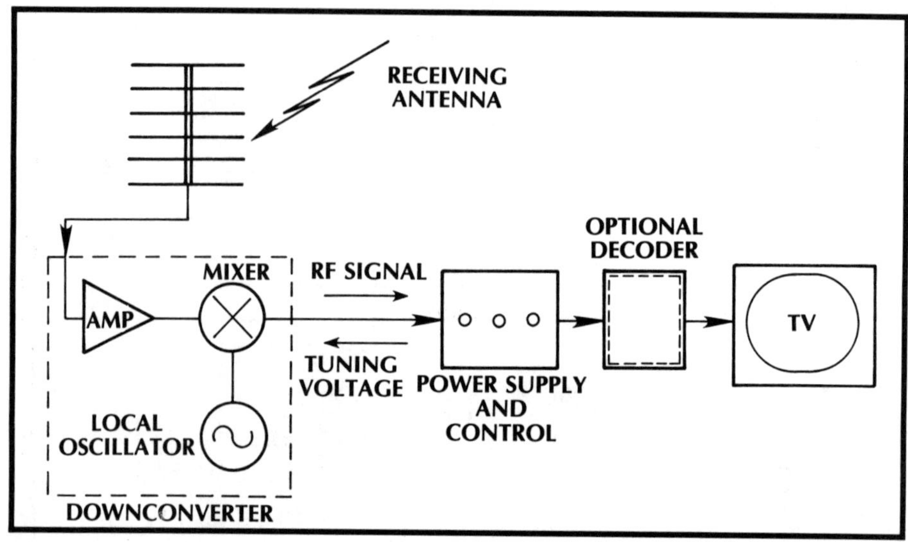

Figure 5-30. MDS Block Diagram. *Signals received by a microwave antenna are amplified and mixed with the appropriate frequency for downconverting and channel selection. The tuning is accomplished by a tuning voltage relayed from the in-door controller to the local oscillator in the downconverter mounted at the antenna.*

The detected microwave signal is channelled into a voltage tuned oscillator (VTO) where it is mixed with the VTO output and is downconverted, lowered in frequency, to a channel 3 IF of 61.25 MHz. Some systems use channel 4 as their target. Since the AM video and FM audio television signal had originally been mixed onto the microwave carrier at the headend, just this one downconversion step yields a TV-ready signal. A final amplification is all that is required. The key to designing a high quality system lies in the frequency stability of the VTO, the core of the downconverter circuitry.

Satellite TV Signal Processing

Signals from satellite TV broadcasts require somewhat more complicated processing. Each channel is relayed by a transponder which has a total FM bandwidth of 36 MHz. These downlinked signals are

Figure 5-31. Conifer MDS Antenna and Downconverter. *The PT-2521 MDS antenna from Conifer has a preinstalled 50 ohm N-connector for attaching to the downconverter. The reflector elements are slotted to reduce wind loading. The MCDU-1600 downconverter is designed for use with all 31 ITFS and OFS channels. The signal from the antenna is attached from a low-loss, N-connector and fed into a 50 ohm jumper cable leading to the downconverter. The output is fed via a F-connector into an RG-6 coaxial cable down the antenna pole to the indoor power supply/tuner. The output frequency ranges from 222 to 408 MHz, from channel K through RR. (Courtesy of Conifer Corporation)*

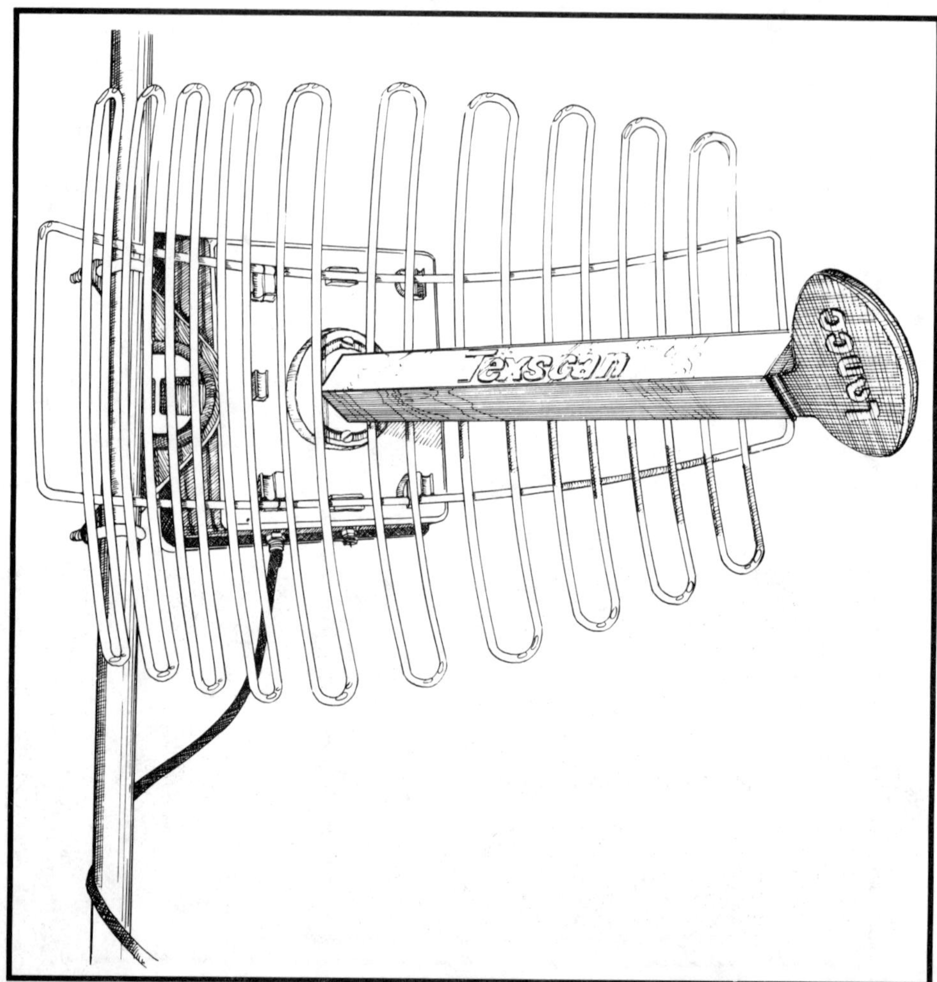

Figure 5-32. Texscan TMC MDS Antenna/Downconverter. *The Texscan has the antenna mounted directly onto the downconverter. Output channels range from 210 to 408 MHz which is compatible with cable ready TV sets and standard set top converters offering in-band scrambling or addressability features. The downconverter includes a microwave pre-amplifier, a mixer and an output amplifier. (Courtesy of Texscan Corporation)*

focused by the receiving antenna onto a collection device called a feedhorn for relay into a low noise amplifier (LNA). The LNA provides approximately 45 decibels of gain while contributing extremely low amounts of noise. The signals are next downconverted and demodulated. Like MDS, channels are selected by varying the mixing voltage sent to a VTO. Unlike MDS processors, the satellite receiver video and audio processors must extract the composite baseband television signal since this raw signal had been FM modulated for effective transmission over the satellite circuit. Finally, the composite audio and video baseband signal is remodulated into a form recognizable by a television, namely AM video and FM audio. (The Home Satellite TV Installation and Troubleshooting Manual by the authors of this book provides much more detail about transmissions and reception of satellite television signals.)

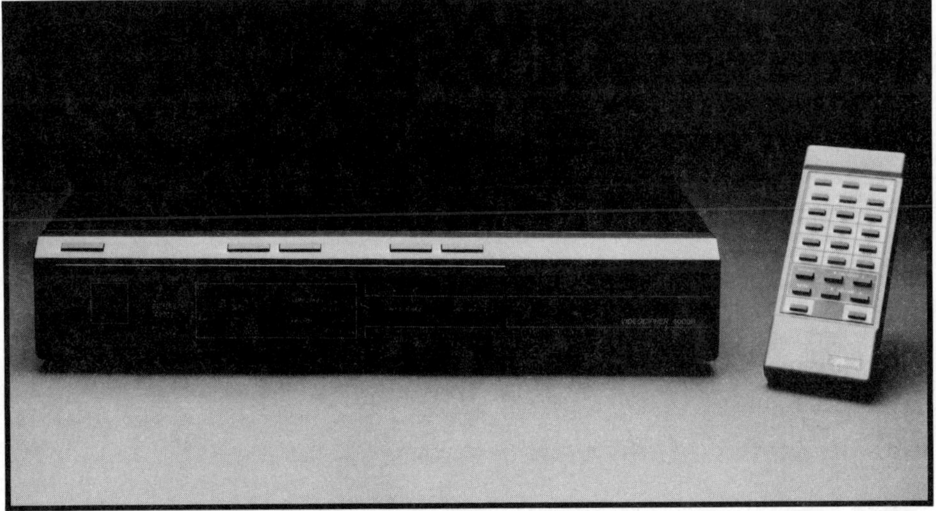

Figure 5-33. VideoCipher IV MMDS. *This decoder used in multipoint distribution services has operational characteristics which closely parallel the VideoCipher II used in satellite broadcasting. (Courtesy of M/A COM)*

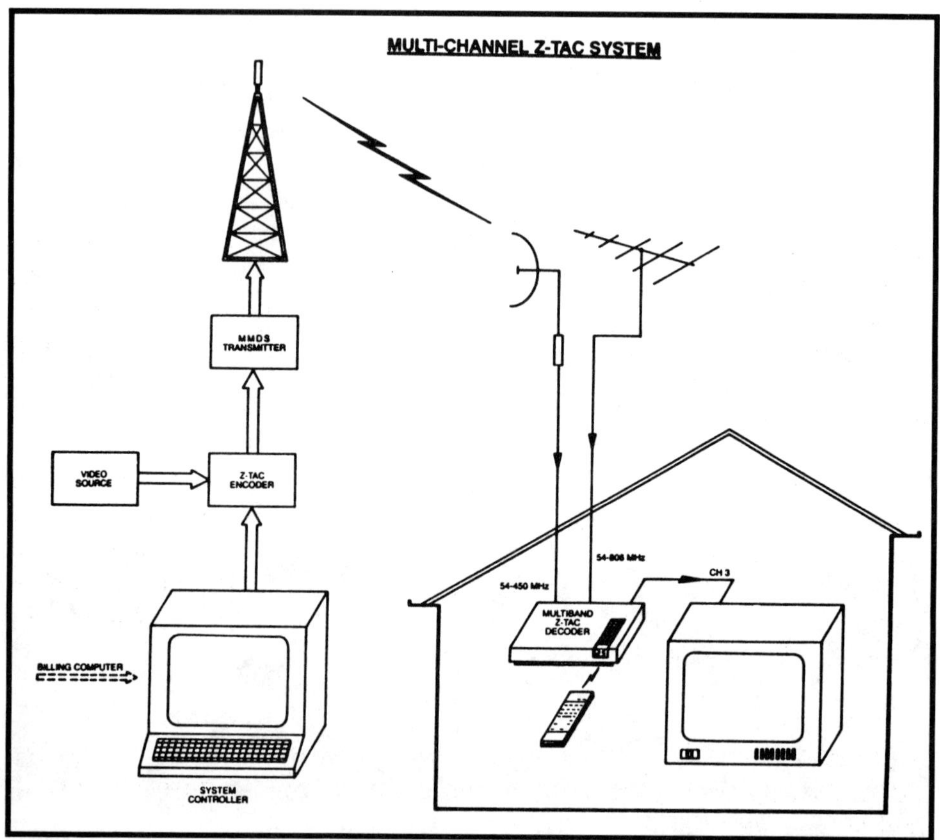

Figure 5-34. Multi-Channel Z-TAC System. *The Z-TAC encoder can also be used to encode and decode MDS broadcasts. (Courtesy of Zenith Electronics Corporation)*

J. RECOGNIZABLE SYMPTOMS OF SCRAMBLING METHODS

There are some telltale signs which allow a trained observer to judge which type of scrambling method is being used by a particular broadcaster. However, the simpler systems have more recognizable symptoms while the more complex may be so distorted that this type of diagnosis will not yield conclusive results. Note that there are other types of audio and video distortion resulting from defects in television receivers. These are not examined here.

When the horizontal sync pulse is either suppressed or removed the scan lines will begin to appear at unpredictable positions and the picture will randomly tear. If an encryption system such as the MAC which causes increasing or decreasing displacement of the start of each successive scan line is used, the picture will pull exhibit a "staircase" or zigzag pattern caused by shuffled lines.

Suppression or removal of the vertical sync pulse will cause the picture to roll. While it may be possible to lock the picture in momentarily by adjusting the vertical-hold controls, the picture will not be stable and soon will fall out of sync.

Effects of an interfering carrier can be quite easily diagnosed. Horizontal bars appear across the screen, the picture rolls and jumps and the sound is distorted. A "tweety bird" noise is usually heard over the audio section.

Sine wave sync suppression causes the horizontal sync pulses to "come and go." Its effect is therefore a vertical black bar with color distorted video on either side which "snakes" its way down the center of the screen. The illumination will tend to be brightest in the center of the picture. The vertical pulses in this and numerous other scrambling systems are not suppressed so the picture retains vertical stability and does not roll.

The gated pulse sync suppression systems all have wiggly vertical bars running down one or both sides of the screen. The SSAVI has video which is intermittently inverted so that colors will be wrong. Skin tones and other scenes with contrasting illumination will take on the appearance of a black and white film negative.

In those cases where the video invert is turned on and off between fields, successive positive and negative images tend to average the illumination out to zero. This is because there is generally only slowly varying illumination changes from one frame to the next. The picture will be severely faded and hardly recognizable.

VI. SATELLITE TV SCRAMBLING SYSTEMS

Satellite TV broadcasts must be encrypted to at least a moderate level of security given their potentially enormous geographical exposure. The challenge is to maintain adequate security while providing decoders at a reasonable cost. High security units for digitally deciphering video signals can cost in excess of $8000. These high costs reflect the fact that digitally encoding and decoding video requires the use of expensive, high speed analog-to-digital and digital-to-analog converters operating at data processing rates as high as 90 megabits per second. In the future, as digital TV becomes commonplace, as lower cost, custom VLSI (very large scale integrated) circuits are introduced, and as companding and other techniques to reduce video bandwidth are incorporated, this type of hard video encoding will become affordable at the consumer level.

Today, most satellite encryption systems rely on digital audio scrambling for high security. The highly encrypted data streams can be embedded within the analog video signal with no concern for interference and crosstalk. And fully addressable features can be incorporated at marginal increases in cost.

A design concern for all encryption systems is in maintaining a high video and audio signal-to-noise ratio during the encoding/decoding process. Methods such as sync suppression ultimately cause an undesirable reduction in television signal- to-noise ratio. Other coding algorithms, as exampled by video inversion, can be either neutral or result in controllable amounts of signal distortion. In contrast, digital methods have the attractive advantage that reconstructed signals can actually have improved quality. Techniques such as error detection and correction, familiar in data transmissions, are at the root of this benefit.

Nine encryption systems which either have been used or proposed for use on satellite broadcast circuits are examined here. The first two, the Satguard and Telease interfering carrier, the latter having been used on the Fantasy channel, are medium security systems not using digital audio encryption. The General Instrument Star-Lok IV was the system of choice for the ill-fated direct broadcast USCI network. Of the remaining six, the Oak Polaris and Link-A-Bit VideoCipher I were designed for high security all-digital audio and digital encryption systems and the MAC transmission system is an entirely new format either scrambled or unscrambled which requires specially designed equipment for reception. The Oak Orion and its upgrade, the Orion-Net, as well as the VideoCipher II are the two systems specifically designed for mass-marketed, direct-to-home satellite broadcasts.

TABLE 6-1. CANDIDATE SATELLITE TV ENCRYPTION SYSTEMS

SUPPLIER	SYSTEM NAME	TYPE OF SCRAMBLING		LEVEL OF SECURITY
		Audio	**Video**	
Comtek	Satguard	Analog	Analog	Moderate
Telease	Interference	Analog	Analog	Moderate
G.I.	Star-Lok IV	Digital	Digital	Very High
Oak/Leitch	Orion	Digital	Analog	High
Orion-Net		Digital	Analog	High
	Polaris	Digital	Digital	Very high
MA/COM	VideoCipher I	Digital	Digital	Very high
	VideoCipher II	Digital	Analog	High
S.A.	MAC	Digital	Analog	Moderate/High

Note: These digital video systems use digital processing, not digital transmission.

A. COMTEK SATGUARD

The Satguard system encodes the video signal with a combination of suppressed horizontal sync, line shifting and pseudo-random video inversion. The advantage of employing the inverted format is that video signal-to-noise ratio is only marginally degraded. The audio signal is securely scrambled by combining it with a complex digital waveform having a frequency at the line scan rate, 15.734 KHz.

The video security is based on the fact that there are 1,000 different line codes available. A data flag embedded in the blanking interval at the beginning of each line determines what form of video scrambling occurs. Line inversion takes place in the middle of the horizontal sync pulse with this pulse being suppressed. As a result, half the picture information is lost and the receiver may not synchronize at either line or frame rates. One large scale integrated (LSI) circuit manages all decoding and enables each of 60,000 units to have a unique address.

All customer decoders have 2 addresses. The common one is used to activate all descramblers while the address unique to each unit must be "refreshed" every 2 minutes or the decoder turns off. Reconstruction of the audio signal is difficult unless both addresses are known and the key is received at the decoder.

B. TELEASE

The Telease is also medium security encryption system. The video is inverted, reduced in level to one half of its normal one volt peak-to-peak and is modulated onto a sine wave of 94 KHz. This sine wave causes the amplitude of the sync pulses in the video signal to be totally unrecognizable. The audio is also modulated onto this 94 KHz signal before being added onto the standard subcarrier, usually at 6.8 MHz.

The sine wave that is superimposed on the video is generated by a crystal controlled oscillator at the uplink. However, no attempt is made to precisely control the oscillator frequency. Since this reference frequency is not transmitted to facilitate reconstruction of the video, the level of security is enhanced. The decoder has further difficulty since the 94 KHz sine wave frequency is near but not exactly 6 times the line scan frequency. Since lines are scanned at 15.734 KHz, the sixth harmonic at 94.4 KHz (6 x 15.734) causes interference or "beats" with the sine wave with an annoying difference frequency of 0.4 KHz or 400 cycles per second.

In order to decode this form of scrambling, the frequency of the reference sine wave must be exactly recreated. This is analogous to using the color burst as a reference to extract the raw color signals or to using the 15 KHz pilot for reconstruction in FM radio receivers. The video signal is split in two components. In one branch the video is inverted and in the other the 94 KHz reference is filtered out so that mixing them together cancels out everything except the original sine wave reference. This can be used to extract the original video and audio signals.

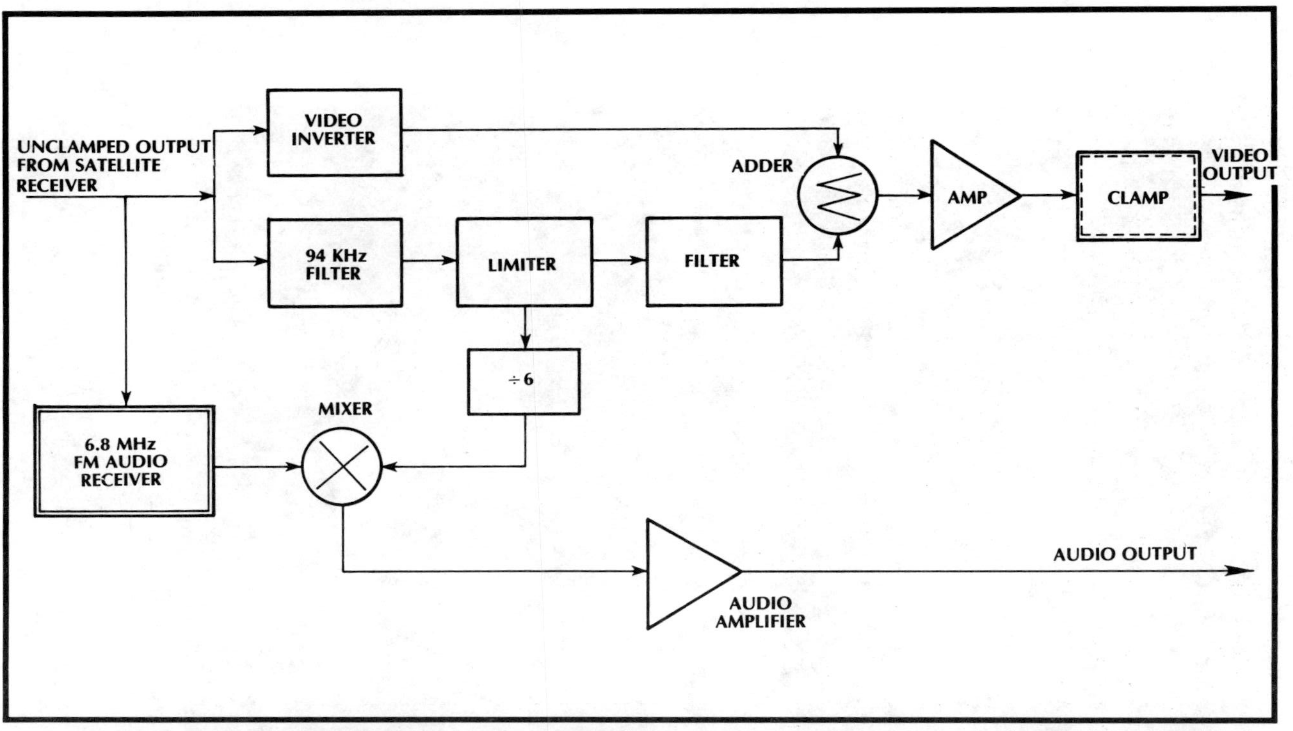

Figure 6-1. Sine Wave Interfering Carrier Decoder. *A 94 KHz sine wave interfering carrier, like the systems used on both FUN and American Extasy Channel scrambling, requires a finely tuned decoder. The original scrambled signal must be inverted and added to the 94 KHz carrier which is extracted from the non-inverted signal to completely eliminate the interference. The audio is also scrambled by modulating it onto a frequency equal to one sixth of the 94 KHz interfering carrier. The required audio demodulation frequency is therefore also recaptured from the scrambled signal.*

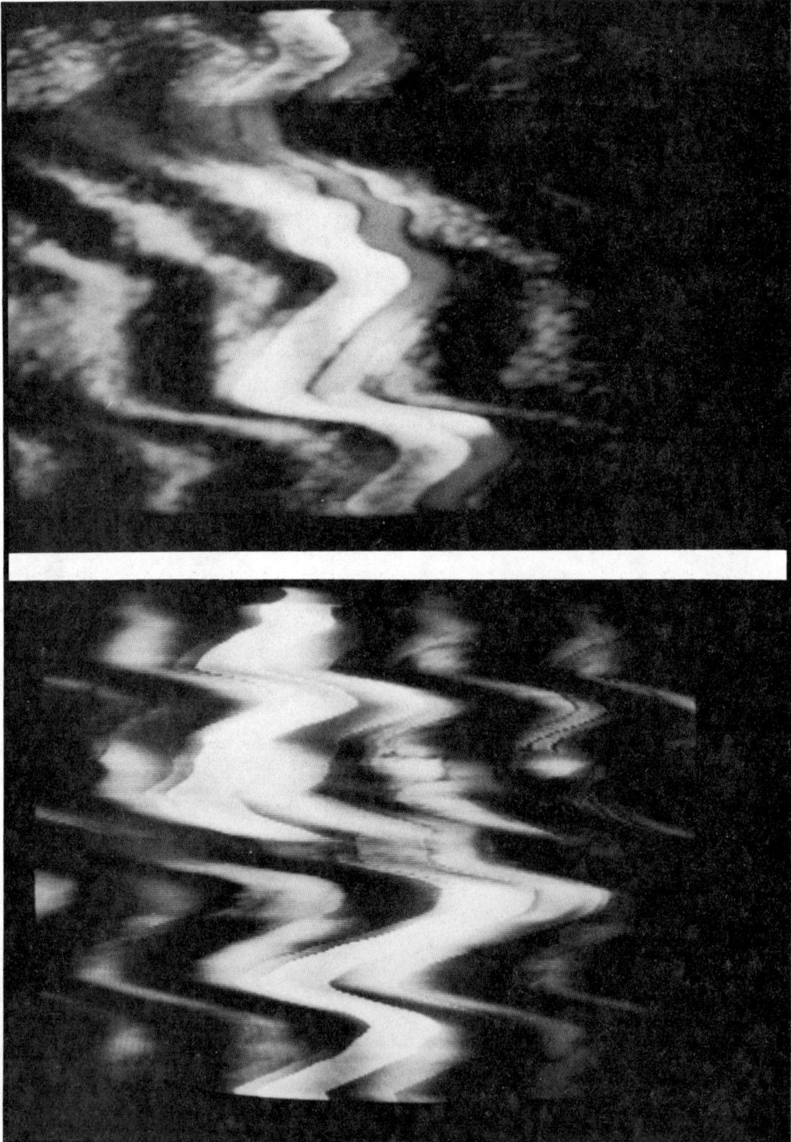

Figure 6-2. Sine Wave Interfering Carrier Scrambling. *Satellite broadcasts from both the Fantasy Unrestricted Network (FUN) and the American Extasy Network use a sine wave of 94 KHz to scramble the video portion of the signal. This single frequency interfering carrier causes the signal voltage to oscillate and, as a result, randomly suppresses the sync pulses.*

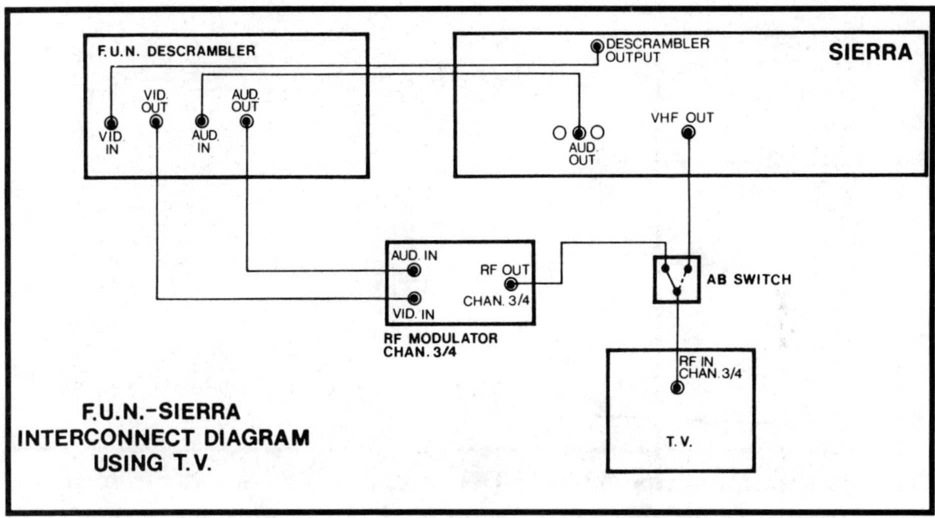

Figure 6-3. Interconnecting the Chaparral Sierra and the FUN Decoder. *The Chaparral Sierra is used as an example here to show how a satellite receiver is interfaced with the FUN decoder. (Courtesy of Chaparral Communications)*

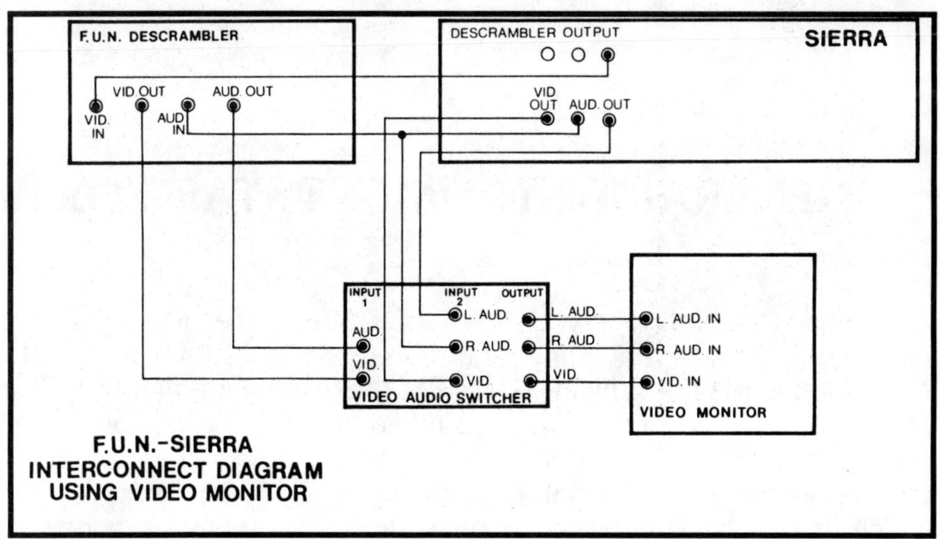

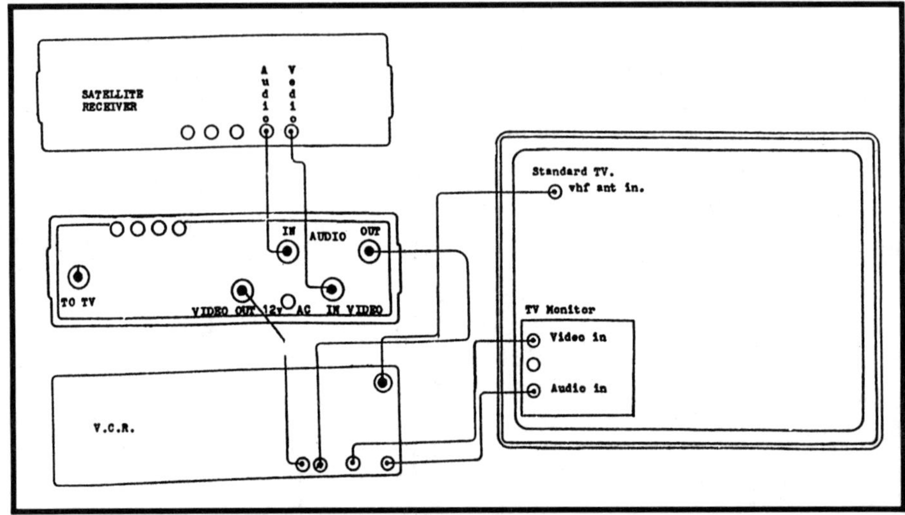

Figure 6-4. Interfacing the FUN Decoder. *The FUN decoder is interfaced with most satellite receivers as indicated in this diagram. (Courtesy of Fantasy Unrestricted Network)*

C. GENERAL INSTRUMENT STAR-LOK IV

The Star-Lok IV encryption system tranforms the video signal into a digital stream for encoding, and then reprocesses it back to analog for transmission. This digital video, like the Oak Polaris and M/A COM VideoCipher I, is not digitally transmitted but only digitally processed. All synchronization and color burst reference pulses are first stripped from the analog video signal. After analog to digital conversion occurs, each line of the video signal is split, diced and rotated according to the system algorithm. Finally, the digital signal is reconverted to analog for transmission. The final picture is totally unrecognizable.

Two digitally encrypted audio channels are available. These are processed with Dolby sound techniques and inserted into the horizontal blanking interval, encrypted and transmitted. Star-Lok IV, like all the other systems using digital audio, uses a forward error correction process. More bits than necessary are transmitted so that error correction circuitry can use the redundant bits to locate and correct any incorrect data.

The system is fully addressable. Up to 16 million subscribers can be addressed at a rate of 750 per second or 2.7 million per hour. A total of 256 tiers are available while up to 32 channels can be relayed per system. One channel contains 1 video, 2 audio and 4 data programs. Addressing can be accomplished on either an individual basis or a group basis. With the group method, subscribers' addresses are embedded into the group code and relayed all at once. Groups can be assembled on a demographic basis so that events that must be "blacked-out" in a particular region can be accomplished virtually instantaneously. Up to 100,000 addresses can be managed in each group.

Like other more advanced encryption and control systems, the Star-Lok IV has attractive customer features. In addition to being capable of providing pay-per-view (PPV), impulse pay-per-view (IPPV) is available. PPV usually means that the subscriber must contact the system operator, usually by phone, to request authorization for reception of special events programming. IPPV can be authorized at the decoder by a touch of a button. The parental lockout feature offers eight levels of control, each one reflecting content ratings as determined by the system operator. A four-digit password controls access.

D. OAK ENCRYPTION SYSTEMS

Presently Oak Industries markets three satellite TV encryption systems: the Orion; the Orion-Net, an upgraded Orion system; and the Polaris, a completely digital, hard security system.

Orion

The ORION (Oak Restricted Information and Operation Network) encryption system uses highly secure audio and moderately secure video encoding.

Video Scrambling

Video scrambling is accomplished by a combination of horizontal and vertical sync removal and randomly timed video inversions. A total of six inversion modes can be selected at will from the control computer. One of four video polarity sequences are used during each frame: alternating with inversion only on even lines; alternating with inversion only on odd lines; all inverted; or all non-inverted. This line polarity sequence is embedded in a digital signal within the horizontal blanking interval of line 22.

The original patent application (#4,353,088) for the Orion system included the possibility of using two additional video scrambling techniques. The voltage level of the digital information in the horizontal interval could be shifted or it could be varied by the application of a sine wave phased to vary the amplitude of the horizontal blanking portion of each line. This latter method is precisely that used by the Oak Sinewave encoder.

Audio Scrambling

Audio is digitized, inserted in the horizontal blanking interval and encrypted for hard security, not by the DES algorithm, but by a 12-digit "word." No audio subcarrier is required. Audio keys can be changed as often as every second. The Orion system, like other satellite broadcast encryption systems, is designed to pass unscrambled audio and video signals through the decoder automatically.

Since the Orion system does not require the use of audio subcarriers, this part of the spectrum is available for use as separate audio or as four 2400 bits/second data channels.

The Orion system is designed to be compatible with conventional satellite transmission and reception equipment and transparent to all the various standard broadcast methods. Thus it can be used on cable systems, MMDS, ITFS or home satellite systems without the need to upgrade equipment or alter system architectures.

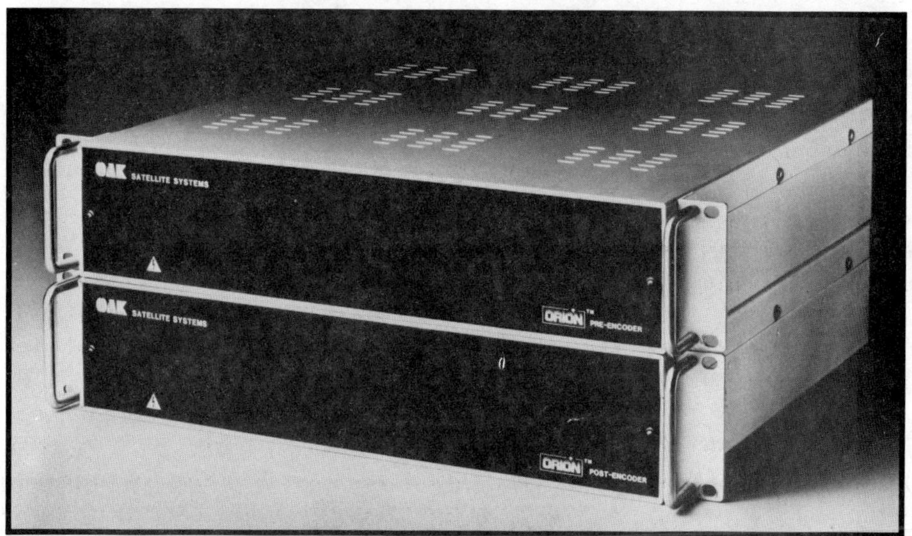

Figure 6-5. Oak Orion Pre-Encoder and Post-Encoder. *These two rack mounted television scramblers convert a standard NTSC signal into an encoded composite video signal with digital audio inserted into the horizontal blanking interval. This equipment can be used either at master or slave uplink stations. (Courtesy of Oak Industries)*

Addressing

All addressing information is transmitted during the vertical blanking interval. The data transmission is structured so that error detection and correction techniques can be used. The actual error rate depends upon the video signal-to-noise ratio entering the decoder as well as on details of the method chosen for digital encoding. Decoders can be addressed at the rate of 1,000 subscribers each minute. However, by preauthorizing decoders, the full subscriber base can be activated virtually instantaneously. The are 49 tiers provided for flexibility of program selection.

Figure 6-6. Orion Database and Addressing Computers. *The choice of either an Apple IIE or DEC PDP 11/44 system control computer determines how many and at what rate subscribers can be managed. The Apple II can address a maximum total capacity of 1,000 subscribers while the PDP 11/44 can control up to 600,000. (Courtesy of Oak Industries)*

System Architecture

The Orion encryption system consists of three principle components: the master uplink; the slave uplink; and the subscriber location. At the master uplink the composite baseband television signal passes through two levels of encoding. The audio is digitized and encoded and the address and control data are inserted in a partially scrambled signal at the pre- encoder. At the post-encoder, the final video encoding is completed and the video and audio signals are combined. This composite baseband scrambled signal then follows the conventional route via the uplink to a broadcast satellite.

The slave uplink is an optional component in the overall system architecture because a customer can receive television signals either directly from the master uplink or from the slave. This secondary uplink adds extra flexibility to the system because it allows the operator to mix in a second program which still contains all the addresses and control data generated at the master station. The procedure to do so is uncomplicated. Once the original message has been decoded at the slave station, the information is applied to a data repeater circuit in the slave uplink pre-encoder. A new audio and video program is inserted at that point before the signal is again processed by an Orion post-encoder.

Note that an additional piece of hardware is usually incorporated into either the master or slave uplink. A device known as a scene change detector adds an extra degree of encoding security by changing modes of video scrambling during intervals when video content changes abruptly. By locking onto average picture illumination levels and inverting the signal at the appropriate times, it can reduce the average power required to transmit the video signal.

System addressability features are determined to some extent by the choice of the control computer. An Apple II series microcomputer used for smaller networks has "menu-driven" software. The higher power Digital Equipment minicomputers permit use of multiple control terminals and transaction/billing control and have special event capabilities as well as menu-driven software among other functions. Note that a total of 2 million addresses are available even though each computer has an upper limit in its ability to manage subscribers.

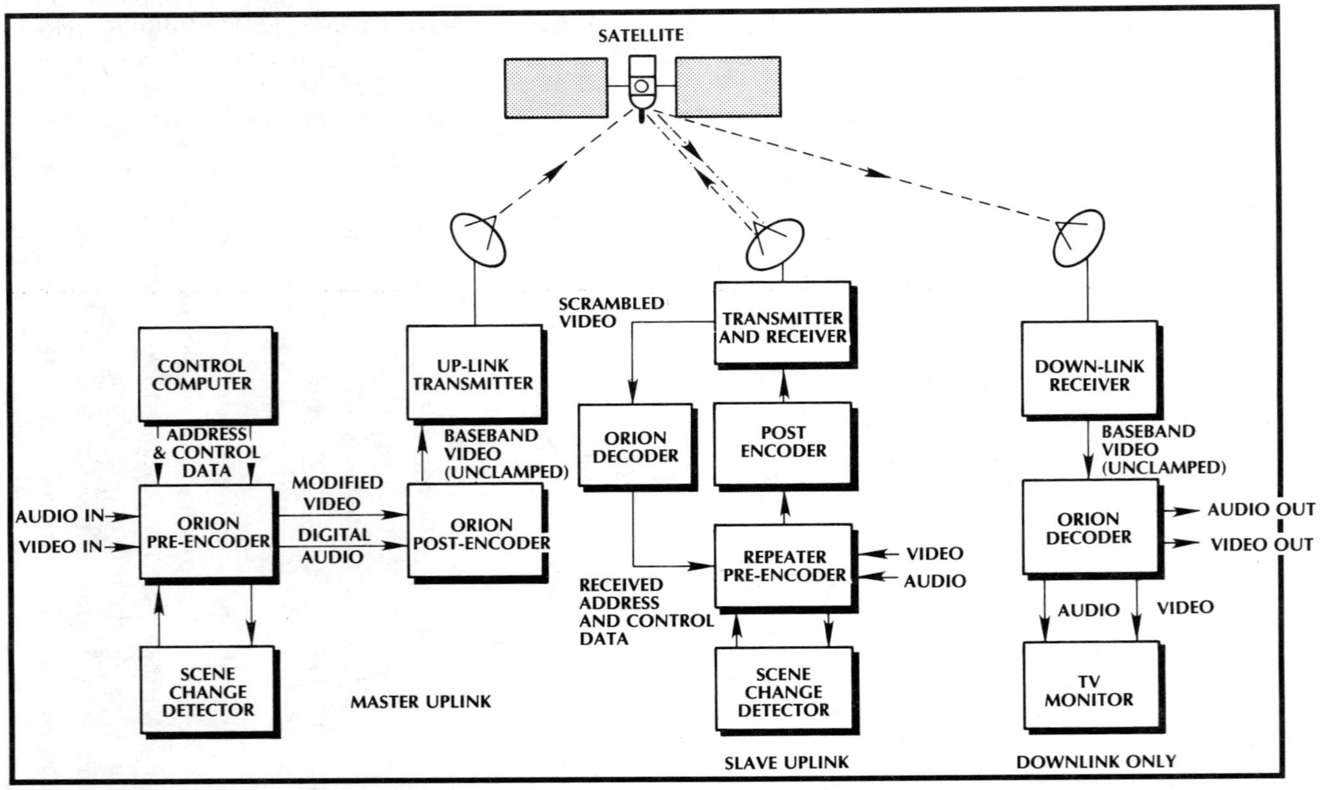

Figure 6-7. Orion Uplink Architecture. *The Oak Orion system is structured so that signals can be uplinked from either a master or a slave uplink with addressing and control data originating from only the master uplink. Adding a scene change detector to the system design improves the scrambling security by adding one more dynamic variable to the overall signal.*

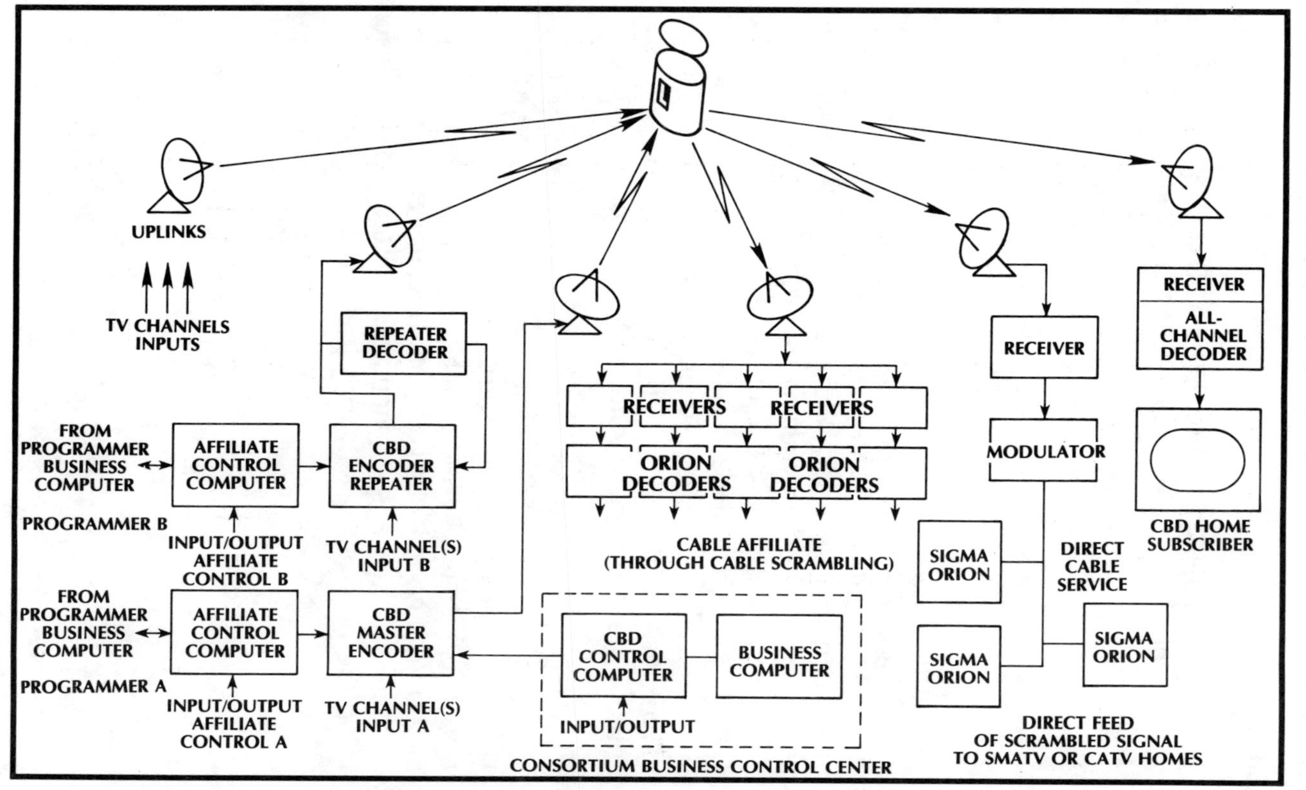

Figure 6-8. Orion Encryption System Architecture. *Broadcasts scrambled by Orion encoders can be decoded by Oak Sigma as well as Oak Orion equipment. A cable TV headend can receive the scrambled signals and remodulate them without decoding onto any channel for relay to subscribers. The Oak Sigma decoder can then demodulate and descramble the broadcast therefore eliminating one decode/encode step.*

TABLE 6-2. OAK ORION CONTROL COMPUTER OPTIONS

CONTROL	ADDRESSING CAPABILITY Typical	Maximum	ADDRESSING RATE/MINUTE	NUMBER OF TIERS
Apple II Series	500	1000	100	49
DEC-PDP Micro 11	5000	30000	1000	49
DEC-PDP 11/44	30000	600000	9000	49

Installation and Activation

Orion decoders come in three different packages: the pro-decoder; the rack mounted personal decoder; and the personal decoder. Oak also makes the personal decoder available in a module for incorporation into satellite receivers by their manufacturers. Installing all three is straightforward. The personal decoder, in addition to having the standard composite baseband audio/video in/out ports, has a television-ready, modulated channel 3/4 output. Both the other two varieties have RS-232 output ports for direct connection to a computer.

The Orion decoder is activated in a two-step process. First, a program level authorization code matching a particular decoder is transmitted. The decoder reads this message and places the information into its memory. This sets the tiering structure of that particular unit. Second, a program level code is relayed with the start of the scrambled audio and video signal. If there is a match between the program level authorization and the program level codes, descrambling begins.

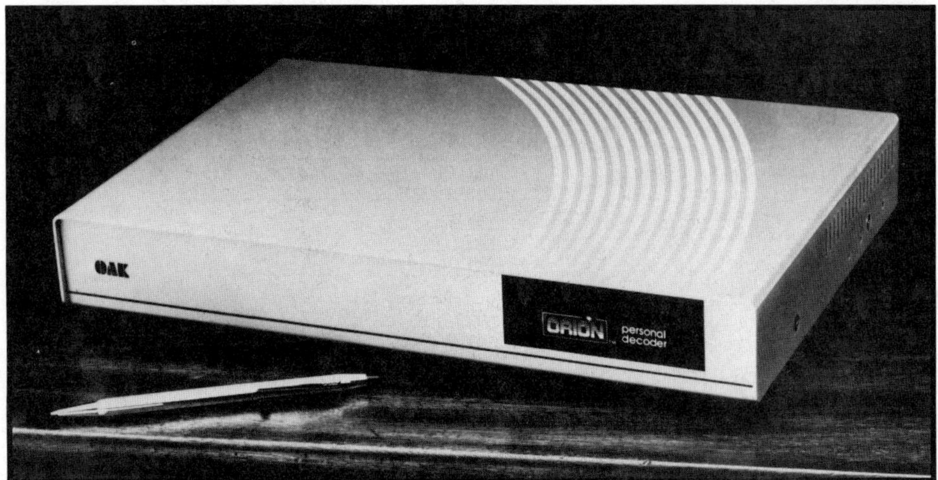

Figure 6-9. Orion-PD Personal Decoder. *This small, set-top decoder is designed for use in direct-to-home applications where low cost is essential. Except for a modulated television signal output to either channel 3 or 4, the features of this decoder match those of all other Orion descramblers. LSI microcircuits decode the digitized audio and data as well as the video information while accepting a variety of receiver video level inputs. (Courtesy of Oak Industries).*

The input specifications for the Oak Orion-PD are as follows:

Video Specifications:
Input Amplitude: 0.5 to 2.0 volts peak-to-peak
Input/Output Impedance: 75 Ohms
Frequency Response: Plus or minus 1 dB from 30 Hz to 4.3 MHz
Signal Waveform: Clamped

Audio Specifications:
Frequency Response: Plus or minus 1 dB from 20-7500 Hz Minus 3 dB at 10 KHz.

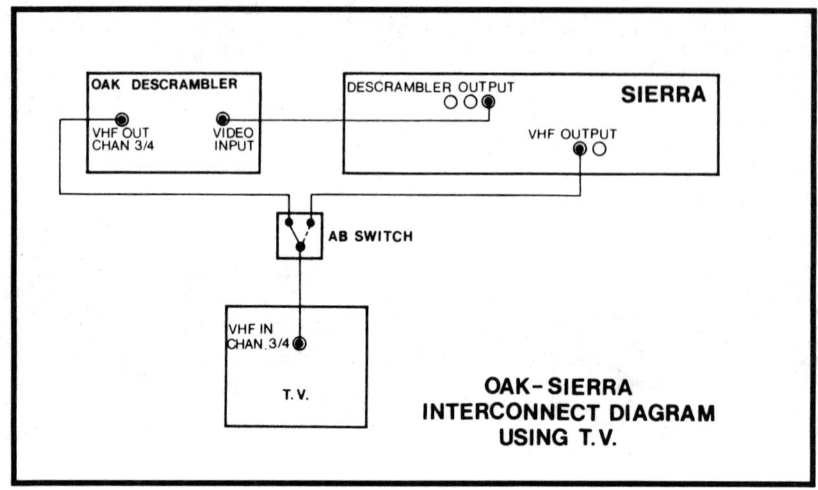

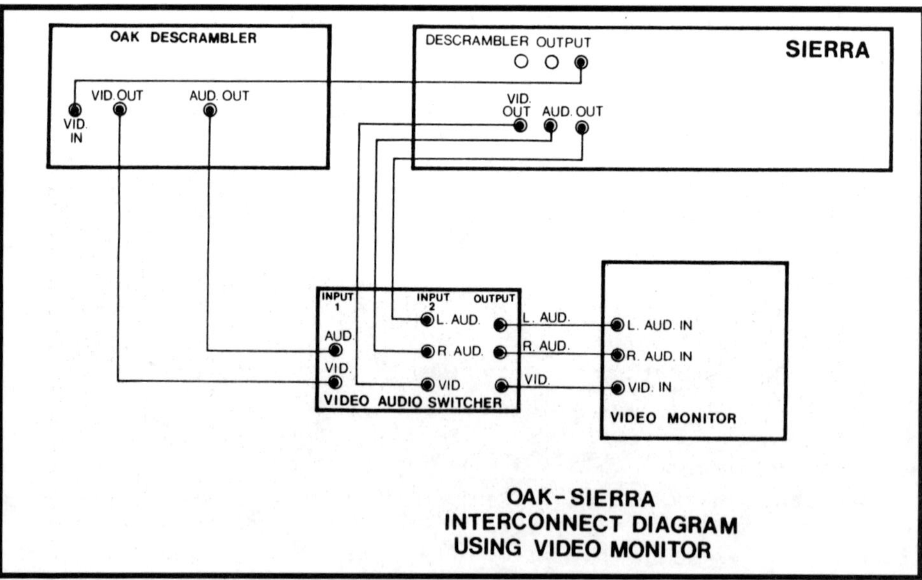

Figure 6-10. Oak-Sierra Interconnecting Diagrams. *This illustrates the interconnection between a Chaparral Sierra satellite receiver and the Oak Orion decoder. (Courtesy of Chaparral Communications)*

Figure 6-11. Orion Rack-Mounted Personal Decoder (RM/PD). *This is the same as the Orion-PD with the exception that it mounts in a 19 inch equipment rack. The only difference in input specifications between the rack mounted and set top decoder is that video input amplitude must range from 0.7 to 1.5 volts peak-to-peak. The audio impedance input and output is 75 ohms unbalanced. (Courtesy of Oak Industries)*

Figure 6-12. Orion Pro-Decoder. *The Orion Pro-Decoder, used in commercial applications, has identical circuitry to the PD decoders but has additional input/output features. This includes three output video jacks, one audio output with both F and BNC connector on the rear panel as well as front panel jacks for monitoring video/audio output. The amount of de-emphasis and filtering can be controlled on the composite baseband input circuits. (Courtesy of Oak Industries)*

Orion-Net

The Orion-Net system is a recent upgrade to the Orion encryption system. It is designed to meet the specific requirements of a multi-channel, direct-from-satellite-to-home broadcast system involving multiple uplinks and downlinks. The video encryption, like the original Orion technique, uses sync removal and random video inversion. Two encrypted audio channels are also transmitted with Dolby digital audio.

Figure 6-13. Oak Orion Scrambling. *The signs of sync removal and video inversion are readily apparent in this picture scrambled by the Oak Orion system. This transmission is from the Canadian Anik D1 satellite tuned to a channel of the Cancom network.*

The architecture has been expanded to allow coordination of independent uplinks owned by different programmers from one Business Control Center. This center has charge of affiliate and individual subscribers and can manage the necessary billing. This architecture is similar to that used by the VideoCipher II system.

Addressability has also been enhanced. Either 49 or 56 program tiers can be managed. Up to 3000 subscribers can be addressed per minute. "Flash" addressing can be achieved at a rate of 78,000 per minute for special pay-per-view events, and incorporating a special option may increase this rate to nearly half a million subscribers per minute. Operators have a choice between systems with 4, 6 or 268 million addresses.

The Orion-Net system is compatible with the Oak Sigma. Encoded signals broadcasted via satellite can be channelled into a small SMATV or cable TV system in which each subscriber uses a Sigma decoder. No intermediate processing is necessary to descramble the broadcast.

Polaris

The Oak Polaris is a hard security digital encryption system for both audio and video signals. It is clearly designed and manufactured for broadcast operators and their affiliates, be they private business or public broadcast networks. It is presently much too expensive to have applicability as a consumer directed system.

Digitized video scan lines are randomly reordered thus rendering the picture totally unrecognizable. The DES algorithm is used in three passes to encode the line order key at an extremely high level of security.

Video test signals which are inserted in the vertical blanking interval are not scrambled and are available on a standard waveform monitor. This is a useful feature for network engineers who can therefore monitor the quality of a video signal without the need to decode the incoming information.

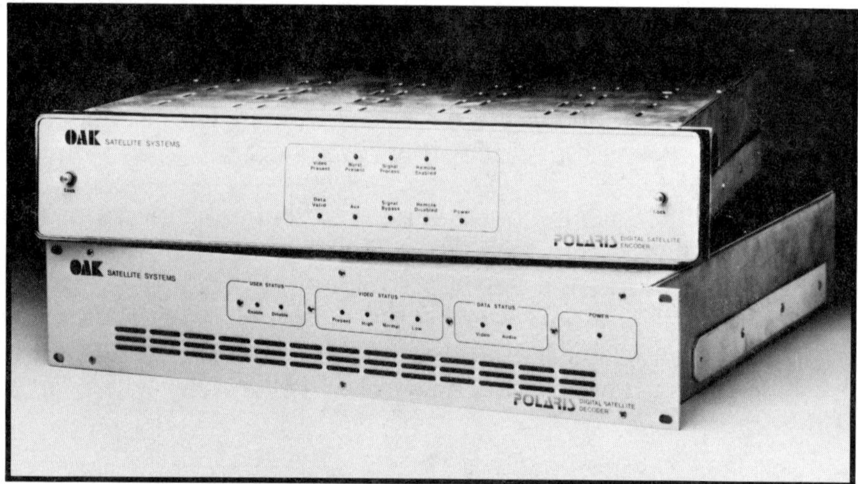

Figure 6-14. Oak Polaris Encoder and Decoder. *The Oak Polaris encryption system is a high level security system which digitally processes both audio and video signals. The video is scrambled by a random ordering of scan lines and the line key is encrypted by three passes of the DES algorithm. (Courtesy of Oak Industries)*

The digitized audio, also encrypted by the DES algorithm, is inserted in the horizontal blanking interval. Up to three audio signals can be multiplexed onto the audio subcarrier before digitizing. The designers claim that the audio quality is comparable to that restored by compact disc players. Error detection and correction techniques are used on both audio and video signals.

The software has 128 tiers and can control 432 decoders. The addressing system is capable of authorizing 14,400 decoders per second, many more than are presently under software control.

TABLE 6-3. USERS OF OAK AND VIDEOCIPHER SYSTEMS

OAK ORION and ORION-NET

Corporate Networks:
- AT&T
- American Medical International
- Chrysler
- Digital Equipment Corporation
- Investment Relations Network
- J.C. Penney
- Merrill Lynch
- Microtel
- Private Satellite Network
- U.S. Army
- Wang Communications

Private Networks:
- American Hospital Association
- Baptists Telenetwork (Spacenet 1/Tr 21)
- Bay Meadows (Westar IV/Tr 6)
- Biznet
- Catholic Telecommunications Network (Westar IV/Tr 9)
- Centex
- Comsat (test system)
- Hawthorn, Pimlico, Gulf Stream (Westar V/Tr 15)
- Hialeah Horseracing (Westar IV/Tr 12)
- Hospital Satellite Network (Spacenet 1/Tr 11)
- Meadowlands & Teleconferencing (Westar IV/Tr11)
- Michigan Hospital Association
- Maywood Park (Spacenet 1/Tr 23)
- Penn National (Westar V/TR 20)
- Pennsylvania Hospital Association
- Philadelphia Park (Satcom IV/Tr 24)
- Pimlico, Garden State Park, Los Alimitos (Westar IV/ Tr 20)
- Santa Anita (Galaxy III/Tr 2)
- Telesat
- Tucson Greyhound Park, GGF Hollywood Park (Galaxy III/Tr 7)
- Voluntary Hospital Association
- Yonkers - Aqueduct & Roosevelt (Westar V/Tr 1)

Entertainment Networks:

Cancom - CHCH (Anik D1/Tr 8)
- WDIV (Anik D1/Tr 9)
- WXYZ (Anik D1/Tr 10)
- TCTV (Anik D1/Tr 14)
- CITV (Anik D1/Tr 17)
- WTVS (Anik D1/Tr 21)
- BCTV (Anik D1/Tr 22)
- WJBK (Anik D1/Tr 23)

French PTT (Secam format)
Philadelphia Park
Prostar
Skychannel (England - PAL format)
Space Age Video
Sugarman Horseracing

OAK POLARIS

American Hospital Supply
Tandem
Voluntary Hospital Association

VIDEOCIPHER I

CBS

VIDEOCIPHER II

Pay Programmers:

HBO
Cinemax
Showtime
The Movie Channel
Disney Channel

Pay-per-view:

Viewers Choice
Request TV
People's Choice

Basic:

CNN
CNN Headline News
MTV
VH-1
Nickelodeon
ESPN
CBN Cable Network
USA Network
WTBS (Satellite Syndicated Systems)
SPN (Satellite Syndicated Systems)
WOR (Eastern Microwave, Inc.)
WGN (United Video, Inc.)
WPIX (United Video, Inc.)
KTVT (United Video, Inc.)

Programmers Considering Scrambling:
- Nashville Network
- Arts and Entertainment Network
- Black Entertainment Network
- Lifetime
- SelecTV
- Event TV

E. M/A COM ENCRYPTION SYSTEMS

M/A COM's subsidiary Link-A-Bit has developed two breeds of scrambling systems: the VideoCipher I and VideoCipher II. The former, like the Oak Polaris system, digitally encrypts both the audio and video for extremely high security. The VideoCipher II is fundamentally the same as the Oak Orion and Orion-Net systems. Although the original Orion did not have stereo audio, the Orion-Net and the VideoCipher II do. The VideoCipher II home unit also has additional customer-controlled operational features which give it an attractive flexibility.

VideoCipher systems are relative newcomers. The Oak Orion equipment has been used in over 20 major systems including over two years of service on the Canadian CANCOM, the US Chamber of Commerce Biznet, and the ON-TV networks, as well as on private broadcast systems of American Hospital Supply, Chrysler Corporation and Merrill Lynch. However, the VideoCipher II has recently occupied center stage having been chosen by networks including HBO/Showtime, Cinemax/The Movie Channel and others. It stands a reasonably good chance of becoming the accepted standard for the lion's share of C-band satellite TV broadcasts.

The VideoCipher system, like the Orion, processes the baseband composite signal and creates a scrambled baseband signal that is concentrated in a 4.2 MHz band in the standard 6 MHz wide television channel. Since the audio subcarriers are removed by being digitized and are embedded in the horizontal blanking interval, the satellite broadcast carrier signal-to-noise ratio can be improved. Manufacturers claim as much as a 2 decibel enhancement.

Figure 6-15. HBO Uplink Facility. *Four 11 meter uplink antennas are used at the HBO uplink facility, the James R. Shepley Communications Center in Hauppauge, New York. (Courtesy of Home Box Office, Inc.)*

Figure 6-16. HBO Scrambling Control Facility. *The HBO control facility at the Shepley center employs a Digital Equipment VAC 11/730 computer. (Courtesy of Home Box Office, Inc.)*

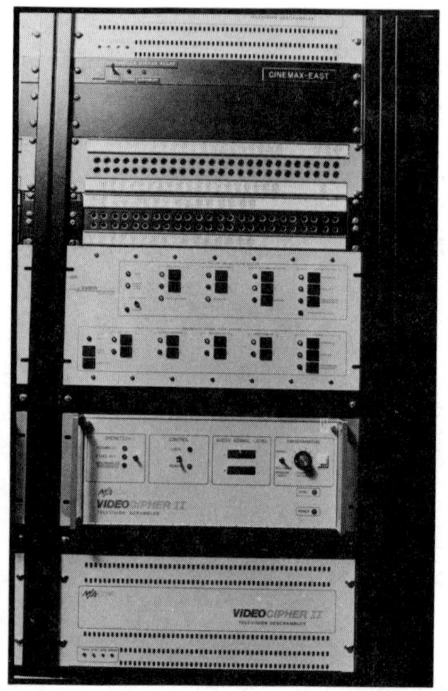

Figure 6-17. Racks of Scrambling Equipment. *Each rack of equipment is dedicated to one satellite circuit or transponder. The overall design includes backup racks for each channel. These include scramblers, descramblers for feeding into television monitors and control circuitry. (Courtesy of Home Box Office, Inc.)*

M/A COM

Figure 6-18. HBO Master Control Center. *Vector scopes for monitoring color phase relationships, waveform monitors, television monitors and other electronic devices are used at the master control center to manage the entire broadcast system. (Courtesy of Home Box Office, Inc.)*

VideoCipher I

The VideoCipher I system uses both hard video and audio encryption. The video signal is processed as follows. Active portions of each line are sampled at four times the color burst frequency. Each sampled point in time is converted to a digital value in a digital-to-analog converter.The higher the sampling rate used in any D/A converter, the finer are the gradations that can be detected as the signal changes, and the more accurate is the digital representation. The digitized line is stored in a multi-line memory and diced according to the DES algorithm into variable sized segments. These segments are then read out in a scrambled order and the resulting digital video is reconverted back to analog form for transmission.

Audio information is sampled at 44.056 KHz rate, a frequency which has a simple relationship to the video frame rate and thus has a low potential for causing interference with the television signal. The resulting 15 bits per sample are compressed and encrypted according to the DES algorithm. This data is inserted into the horizontal blanking interval along with control data.

Digital audio enables use of error correction techniques which detect and correct all single bit errors as well as detect and conceal all double bit errors by interpolation. The DES reshuffling of the input bits causes any burst errors resulting from the detection of bursts of noise during transmission to be received as less destructive random independent errors.

Unlike the digitized video signal, the audio digital rate is low enough so that scrambling can be accomplished on a bit-by-bit basis. The respective bit and line shuffling or scrambling for the audio and video signals is encrypted by the DES algorithm using a 64-bit key. This 64 digit key translates into a possible 2 to the 64th power number of combinations, 1.84 times 10 to the 19th power or about 18 billion billion combinations.

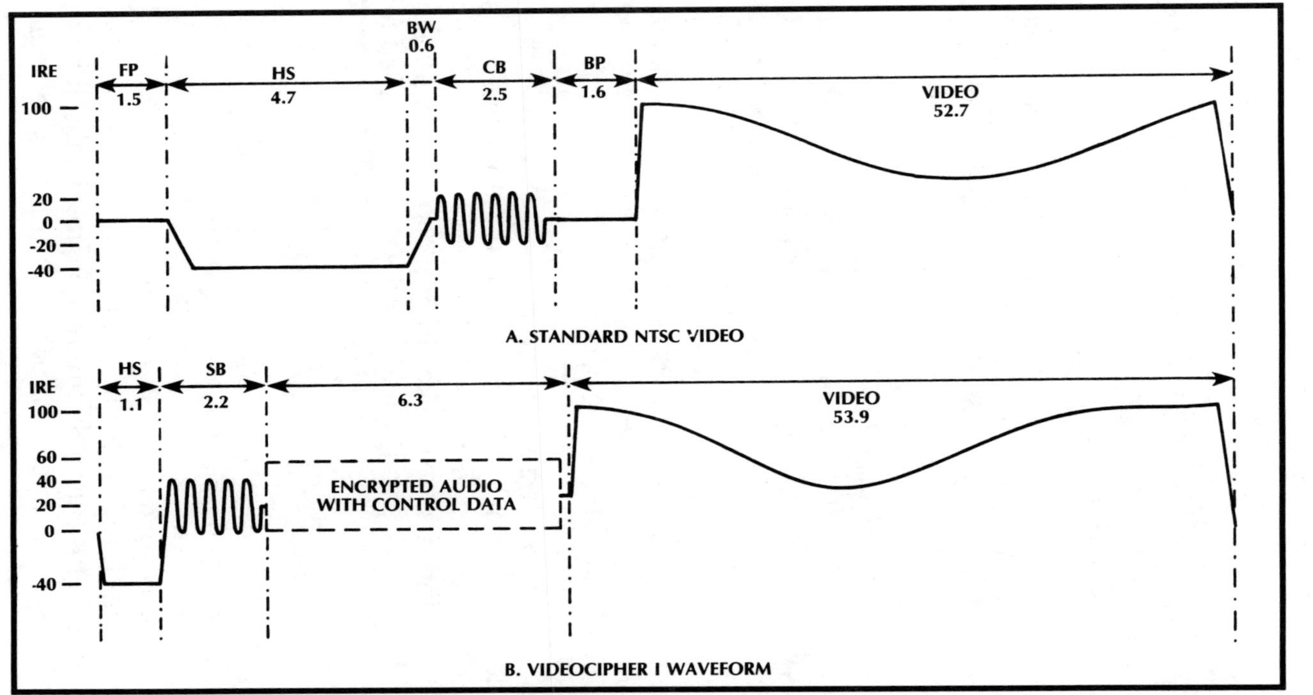

Figure 6-19. Video Waveform Before and After VideoCipher I Encryption. *This illustration shows the changes in video waveform that occur in encryption with the M/A COM VideoCipher I. The numbers are time in microseconds. The legend is as follows:*

BP - Back Porch
BW - Breeze Way
CB - Color Burst
FP - Front Porch
HS - Horizontal Sync
SB - Sync Burst

Addressing and Control

The control architecture is designed so that (1) individual decoders can be addressed for authorization as well as to establish control categories, (2) control categories can be addressed by one command, and (3) the entire decoder population can simultaneously receive broadcasts of control information.

Key distribution and security have been carefully designed and are also arranged in a hierarchical fashion. Each decoder effectively contains a large number of pre-determined keys, each of which is selected at the uplink by a "key generation number." Therefore, any one uplink unit key list corresponds to just one key generation number. If the list is stolen, it can be replaced by activating all decoders with a new key generation number. In addition, keys are changed at regular intervals, for example every half hour or at the beginning and end of each program. Keys are securely locked into a single-chip Intel 8751 microprocessor.

Channel keys are established to unlock specific channels or programs. These are distributed on a category basis so that a large population of decoders can be authorized at once.

All the control data is processed according to an error detection technique at the decoder. If an error is detected the transmission is rejected. In order to further minimize the possiblity of a data reading error occurring, any message conveying key information is periodically retransmitted and must be received twice in order to be accepted as accurate.

Control and Management Computers

Control and management computers provide two functions: database management and real-time control message generation. A DEC VAX 11/730 computer is primarily responsible for database functions while an IBM PC or other microcomputer controls the user interface.

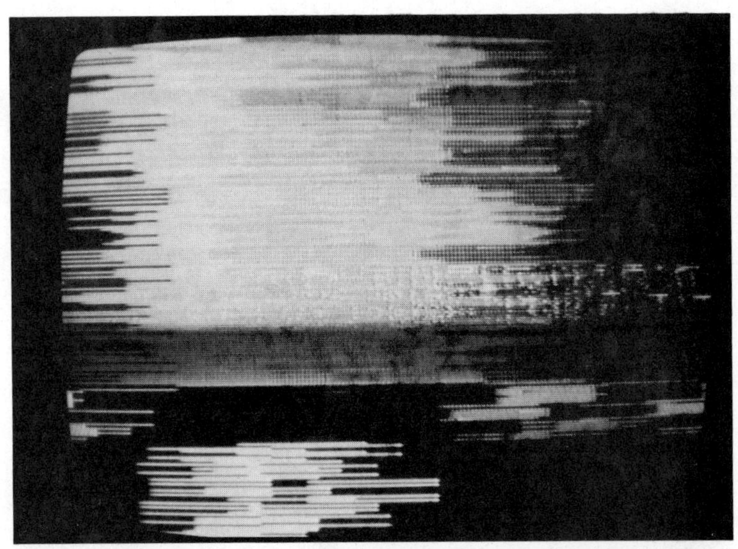

Figure 6-20. VideoCipher I Descrambling - Before and After. *Video encryption with VideoCipher I is hard and the picture is barely recognizable before decoding. The black region across the color bar pattern is an artifact due to lack of synchronization between the field scanning and 35mm camera shutter.*

VideoCipher II

The VideoCipher II encryption system accepts a standard NTSC video input, two audio inputs and an auxiliary data channel at baseband. The scrambler digitally encodes the audio signal for hard security and uses conventional analog encryption techniques to achieve a moderate level of video security. The scrambled signal is contained within a standard NTSC bandwidth and can be combined with additional subcarriers and transmitted over any type of satellite television link.

Clear video, three audio outputs (left, right and monophonic) and the auxiliary data signal are decoded by a commercial unit at a cable TV headend. The baseband descrambled signals are modulated and possibly re-scrambled for distribution to subscribers. Some systems such as the Oak Orion need not decode the signal before transmission over cable lines because Oak Sigma decoders can descramble a modulated signal.

The consumer VideoCipher II produces baseband video for a monitor, two audio outputs for a stereo and a remodulated channel 3 or 4 video and monophonic audio output for a TV set. An auxiliary data output is also provided for downloading software and for controlling various devices at rates up to 88 kilobits per second.

Video Scrambling

The composite baseband video signal is scrambled by eliminating both the horizontal and vertical sync pulses, by inverting the video waveform and by centering the color burst at a non-standard level. This modified color burst is suppressed to the black level to prevent newer digital televisions which can easily regenerate sync pulses from finding it. Although it is in the correct temporal position relative to the video signal, it is unlikely to trigger the horizontal synchronization circuits.

Altogether, 13 combinations of video scrambling are possible. In the decoder, custom designed VLSI chips detect the digital sync pattern in each frame in order to reconstruct the original sync pulses. These chips in combination with other discrete circuit elements are also used to restore all components of the original baseband video signal.

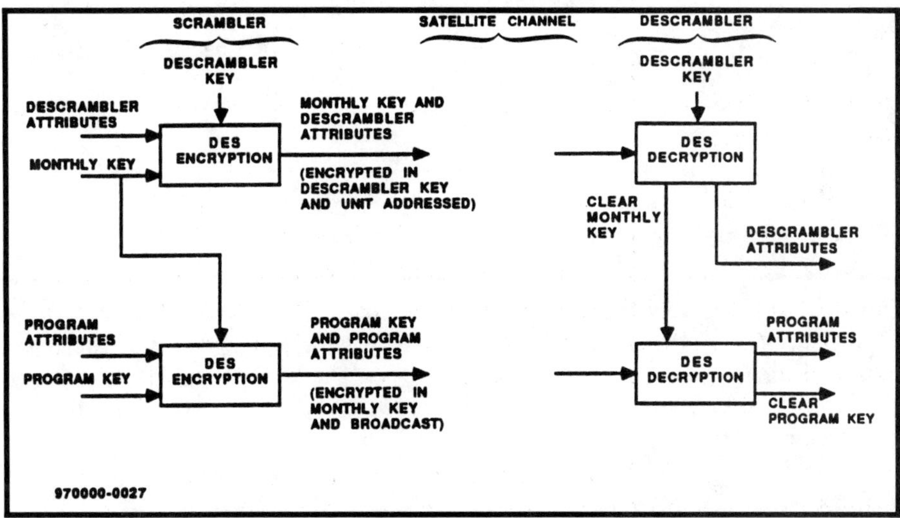

Figure 6-21. VideoCipher II Key Hierarchy and Distribution. *Both the monthly key and program key are uplinked to every VideoCipher II decoder. Tiering structure and other program information can be encrypted into the monthly key. The descrambler attributes refer to the specific decoder address. (Courtesy of M/A COM)*

Audio Scrambling

The two digitized stereo quality audio channels combined with the addressing and control information and auxiliary data channel are inserted in place of the horizontal sync pulse in each scan line as three 8-bit digital words. Removing the audio subcarrier allows higher power to be applied to the video signal so that video performance can potentially be improved. The DES algorithm, expressed as a 56-bit key or password, encodes the total digital signal. Using a 56-bit key translates into 2 to the 56th power or over 72 quadrillion possible keys.

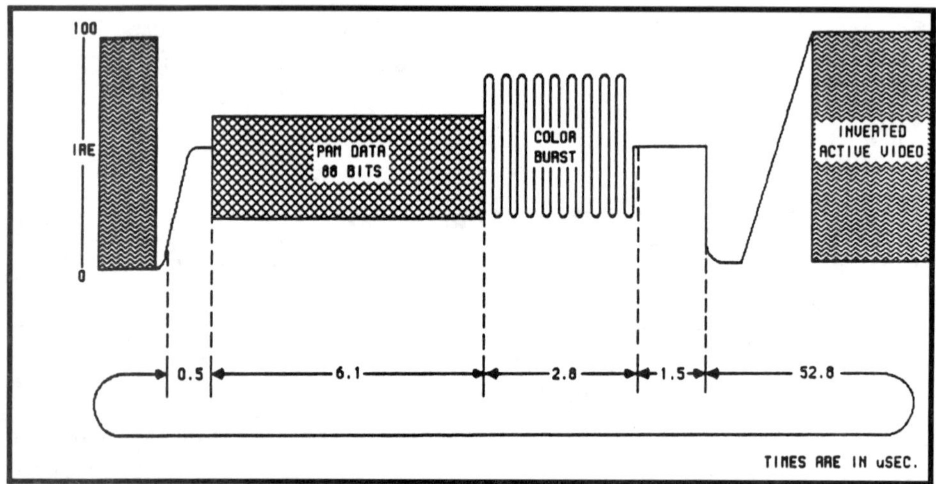

Figure 6-22. VideoCipher II Horizontal Line Format. *Each line of video information is encrypted as shown in this illustration. Time is expressed in microseconds. PAM data refers to a type of modulation known as pulse amplitude modulation. The active picture portions of the signal are inverted. (Courtesy of M/A COM)*

Addressing

Each VideoCipher II decoder has a set of unique keys, expressed as a string of 56 ones and zeros, stored securely in a non-volatile memory. A non-volatile memory will not be erased during a power failure or when the decoder is unplugged. This DES key is the crucial element in descrambling a broadcast.

This program key, which can be changed once every 30 minutes or at program boundaries, is uplinked whenever a customer is activated or whenever tiering or billing information is altered. It is equivalent to a telephone number. The correct number must be dialed before a connection is made and before any type of information is transmitted. However, unlike a telephone circuit which remains connected until one party hangs up, the VideoCipher II has an internal clock that will deauthorize after a prescribed time even if the input to the decoder happened to have been removed. Also, each decoder has many program keys or telephone numbers for further security.

These program keys can all be embedded into a more general key known as a monthly key. This name is chosen since most billing occurs on a monthly cycle. The net result is that all authorized decoders can be turned on by a single transmission each month or at any selected time period. The details of this addressing system is similar to that used in the VideoCipher I system.

The tiering structure allows one or any combination of 56 programs to be delivered to any particular decoder. A tier can be general enough to include all 56 programs or be as specific to only include one pay-per-view broadcast on a particular channel. Tiering information is transmitted with the monthly key.

The tiering structure can also be coordinated with market information such as customers' zip codes. This allows decoders in a particular region to be simultaneously blacked out or activated by a single command. This ability will prove useful to the evolving breed of "program brokers." They will purchase programming at discount rates for many subscribers at once from different broadcasters for packaging and resale nationally.

Of course, authorization and tiering information can be relayed at any time to activate or deactivate individual decoders or to change tiering information. The maximum potential addressing rate is 600,000 subscribers per hour or about 167 per second. This system is structured to be able to manage about 100 million decoders, more than enough for the entire North American continent.

System Architecture

The VideoCipher II architecture is simpler and less flexible than that used in the Oak Orion/Orion-Net systems. These are designed so that slave as well as master uplinks can control all addressing and program origination. The VideoCipher II addresses all the decoders from one central computer at the Link-A-Bit facility in California. While decreasing system flexibility, this structure has the advantage that broadcasters need not own and operate a sophisticated network control center.

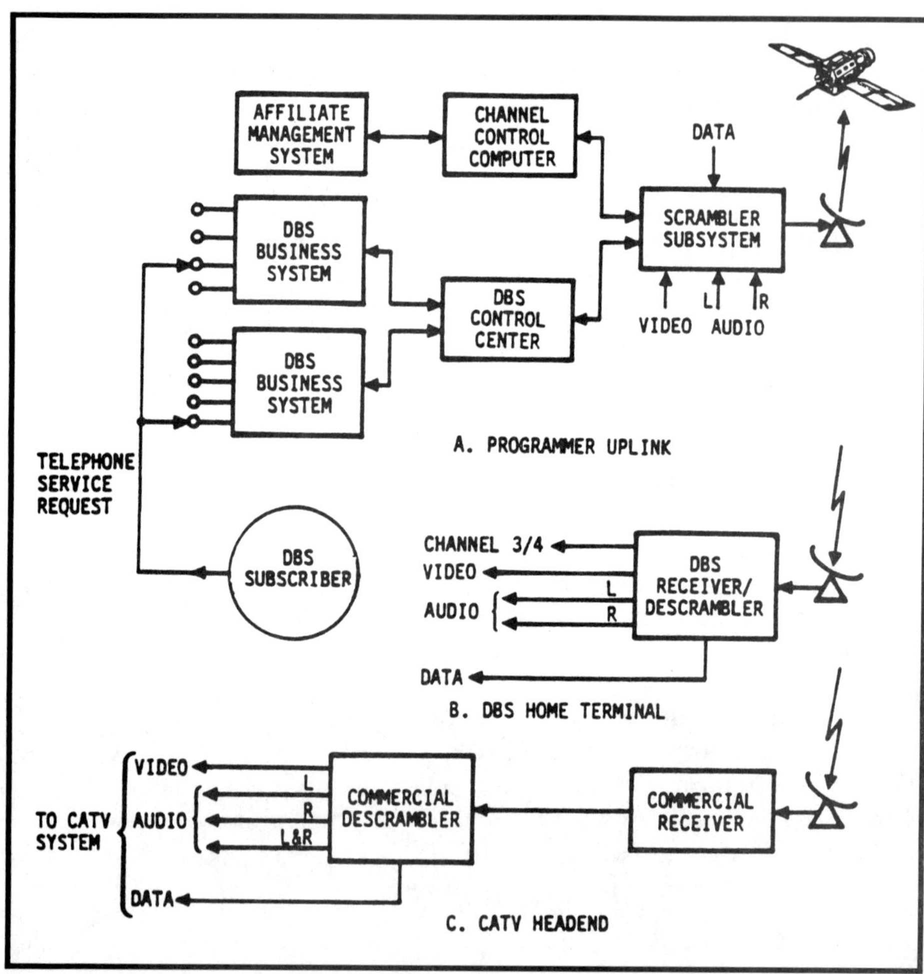

Figure 6-23. VideoCipher II Satellite Television Scrambling Architecture. *The overall VideoCipher II architecture permits reception of scrambled broadcasts by both individuals and cable TV networks. Individual subscribers can request service from either the control center or cable TV office. (Courtesy of M/A COM)*

Figure 6-24. VideoCipher II Descramblers During Testing. *Before scrambling was implemented many racks of VideoCipher II units underwent testing. (Courtesy of Home Box Office, Inc.)*

At the Link-A-Bit control center, four redundant high speed mainframes communicate via land lines with each broadcaster's computer. The addressing information is added to the scrambled video signals at their uplinks. The central computer facility controls subscriber access to scrambled signals. Up to twenty-four different business system computers can interact with this center. However, subscriber data is partitioned in the central computer so that information specific to any one operator is unaccessible to all others. The business computer contains the subscriber database, the descrambler unit address, all subscriber billing and service information and the subscriber authorization code. Customers interact directly with personnel at the central facility.

Up to 240 different satellite channels can be handled by the central uplink facility. These can be packaged in any combination onto any of the 56 available tiers. Programs can be simultaneously allocated to any particular tier.

Figure 6-25. The VideoCipher II Consumer Decoder. *This is the decoder whose makers expect will transform the home satellite business. It is available as either the baseband/RF model 2000 E or the baseband-only model 2000 E/B. The level of the RF signal must fall between -55 dBm and 0 dBm (3 microvolts and 1 millivolt). The baseband input should be close to 1 volt. Other rigorous input specifications concerning frequency response and the balancing of chroma and luminance signals have been published by M/A COM. A minimum signal-to-noise ratio of 9 decibels is also recommended. This implies that pictures with more than just a few sparklies may have too low a ratio to function properly. The VideoCipher II has a shortcoming in that it overrides satellite receiver volume controls on encrypted channels and allows no adjustment from the receiver's remote control. The VideoCipher is also available as a module which can be built into satellite receivers. (Courtesy of M/A COM)*

Broadcasts can be formated in four ways:
(1) The standard mode is fully scrambled and uncoded.
(2) Fully scrambled and coded. An error correction technique can be incorporated to minimize problems with subscribers who have difficulty being authorized. Use of this method has the disadvantage that the number of subscribers who can be served by the system is reduced.
(3) Fixed key mode. In this mode an encrypted signal is transmitted but all decoders are capable of descrambling the information. This could be used on a promotional channel and is useful in system testing.
(4) Only the video can be scrambled while the audio is relayed in the clear.

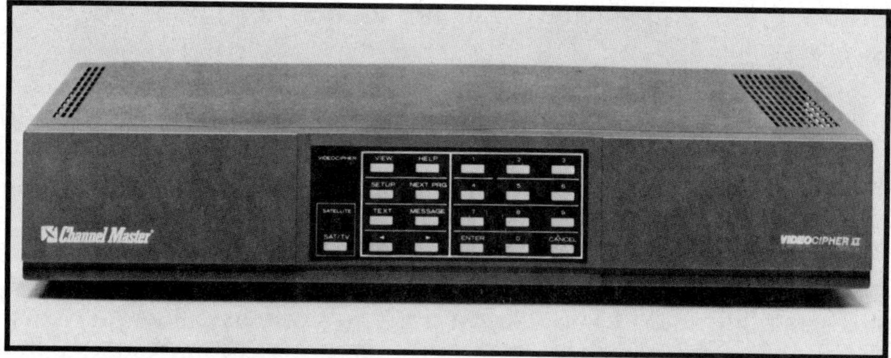

Figure 6-26. The Channel Master VideoCipher II. *Channel Master private labels this technology licensed from M/A COM Link-A-Bit. No on-screen graphics from the satellite receiver are passed through the VideoCipher II and displayed when viewing scrambled channels. Only the VideCipher II on-screen graphics can be viewed. (Courtesy of Channel Master)*

Customer Interfaces

The VideoCipher II has the attractive capability of being able to generate text and numbers called alphanumeric displays on screen via a built-in character generator. This display is nine lines of 20 large characters each and is superimposed over the picture. This feature can be a very useful tool for customer feedback and interaction.

There are three sources for on-screen messages:

- Messages stored with the decoder's read-only memory with text such as "SORRY, BUT YOU ARE NOT AUTHORIZED TO RECEIVE THIS PROGRAM" or "BLACKOUT IN EFFECT."

- Messages of any sort transmitted via satellite.

- Numbers entered by the subscriber used, for example, in parental lock-out.

This alphanumeric display can be coupled with a range of useful features including electronic mail, pre-paid and impulse pay-per-view, selection of a second language, diagnostics and other on-screen status messages.

The electronic mail capability allows a programmer to relay any type of message such as stock market quotes to just one decoder or to any selected group in its customer base. Up to 256 pages of text organized into any number of "scrolls" can be transmitted on each channel. It is perfectly within the capabilities of the system to be able to have one customer pay to have a birthday message or other greeting sent via satellite so as to appear on a friend's television set anywhere in the country. And all these messages are stored in the VideoCipher II memory until the "Message" key on the unit's front panel is activated.

The prepaid pay-per-view capability allows a customer to receive any type of programming for a specified period. Impulse pay-per-view (IPPV) is built around a "credit register" which is accessed by a subscriber via a user-defined password. When IPPV payments are made, the control center uplinks a message and increments the credit register by a specified amount. The on-screen display shows the IPPV program name, its standard movie industry rating when applicable, the cost, credit available to the customer and time remaining in the program. Subscribers can even lock-out movies of undesirable ratings. Each time an IPPV movie or special event is selected, the credit register accounts for the expenditure. When the credit limit is reached, a message indicates that no further programs are available until the account is advanced.

A second language can be transmitted over one of the two stereo channels. When this is done, authorized customers can select to hear either language.

In addition, a wide range of alternative on-screen displays are possible. For example, whenever a new program begins or the channel is changed, the name and time remaining can be automatically displayed. Whenever a problem arises, the VideoCipher II which has built-in, on-screen diagnostics can display a message prompting the subscriber to take corrective action such as calling the toll-free Link-A-

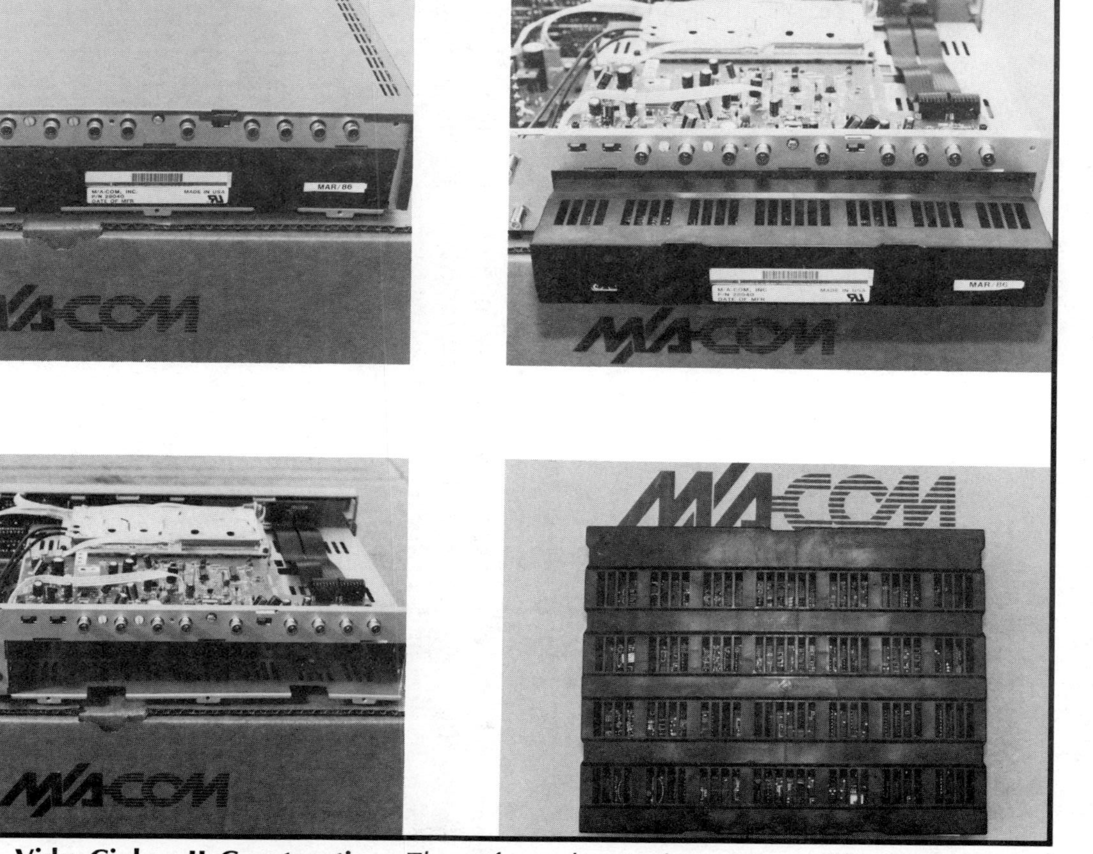

Figure 6-27. VideoCipher II Construction. *These four photos show where the VideoCipher II descrambling module is located. The unit in photo A is enclosed. The module has been removed in photo C leaving what is essentially a satellite receiver without a tuner section. Its power supply is located in the left-hand portion of the chassis. When units are assembled at the factory, the identity of each VideoCipher II is established by the serial number and programming of the descrambler module.*

Bit center for help. Twenty-eight on-screen "assist" displays are also available to aid the consumer in operating the various functions.

Installation and Activation

Hooking up a VideoCipher II to any satellite receiver is a rather simple job. Some of pages from the thorough M/A COM installation guide are reproduced here. Once the electrical connections are completed, these simple steps are followed:

(1) When using a 2000E, set the rear switch to either 70 MHz or baseband input.

(2) Tune in one of the satellite channels scrambled by the M/A COM system. The screen should go completely black. The front panel light indicating reception of a scrambled signal should be lit. If not, or if the screen still displays scrambled video, recheck the connections. Adjusting the satellite receiver input level control may solve the problem. In an environment with terrestrial microwave interference, filters may have to be used to reduce it to a level substantially below that of the video signal (i.e. lower than at least 7 or 8 decibels. Please see "The Home Satellite Installation and Troubleshooting Manual" for more details.)

(3) Press the setup button. The 12 letter and number unit serial number should appear. There will also be an on-screen numerical display of the signal strength reaching the decoder. This ranges from zero to a maximum of 50, with 45 and above considered acceptable. Input levels below 40 could result in garbling of messages and unacceptable audio and video quality.

(4) Call the chosen programmer or local cable company, usually via a toll-free number. The 12 digit serial number and, most often, a credit card number will be required. Once the operator enters the decoder's serial number and desired program selection or tiering structure, the computer software searches for the program keys and transmits them over the appropriate satellite channels. The decoder must remain tuned to the scrambled service until a clear picture snaps onto the screen. Whenever this channel is selected at future times the screen

will be momentarily black, but is soon followed by a clear picture. The second light indicating operation of the VideoCipher II should also be lit. Whenever a clear signal is received, this indicator light will remain off and the decoder will be bypassed.

(5) Finally, balance the video and audio signals between scrambled and clear channels. This is accomplished via rear panel, screw adjustments.

When authorized on scrambled channels, the decoder produces a clear composite baseband video signal, a modulated channel 3/4 video output and left and right stereo channels. An auxiliary data output with capability for transmitting up to 88 thousand bits per second is also available for downloading software or controlling various devices. Even if the satellite receiver in use is not equipped for stereo, the outputs from the VideoCipher can drive a stereo on only encrypted channels. Note that most of the decoder features such as teletext are not being used on present scrambled satellite channels.

M/A COM

Figure 6-28. VideoCipher II Circuit Board. *The critical components of the descrambler are shown in the board layout. The key is as follows:*

U1 - Texas Instruments (TI) - TLO 71 CP
U2 - AMI - 8545 MBA/11627
U3 - SOCN - 24766-1
U4 - AMI - 8516 MAY/11627 SOCN - 24298-1
U5 - NCR - 0380343 SOCN - 11627 4*M* - 24297 A8547 - F818981
U6 - Fugitsu - MB88303 / 8445-EOO
U7 - TI - TMS-7000 - SOCN 24294-3 - C11202N MAS-8522
U8 - RCA - CD4053BE/ H-419
U9 - (Motorola) - SN 74-LS-125-AN
U10- TI - SN7406
U11- TI - 74LS-374
U12- TI - 74LS-244N
U13- SD-5000N
U14- TI - TLO-72CP
U15- TI - SN74LS-139 AN
U16- TI - SN74LS-373 N
U17- TI - TLO-72 CP
U18- LN3362
U19- TI - TMS 7001 NL4
U20- Toshiba - TC5516APL-2
U21- RCA - CA-3080 E
U22- RCA - CA-3080 E
U23- NE-521N
U24- NCR - 0380342
U25- not used
U26- Harris - A1-2540-5/8446
U27- NE-521N
U28- TI - SN74AS-174N
U29- RCA - CA-3046
U30- 27128A-2
U31- F - 74F86
U32- TI - TLO71CP
U33- RCA - CD4066BE
U34- TI - SN 74AS74N

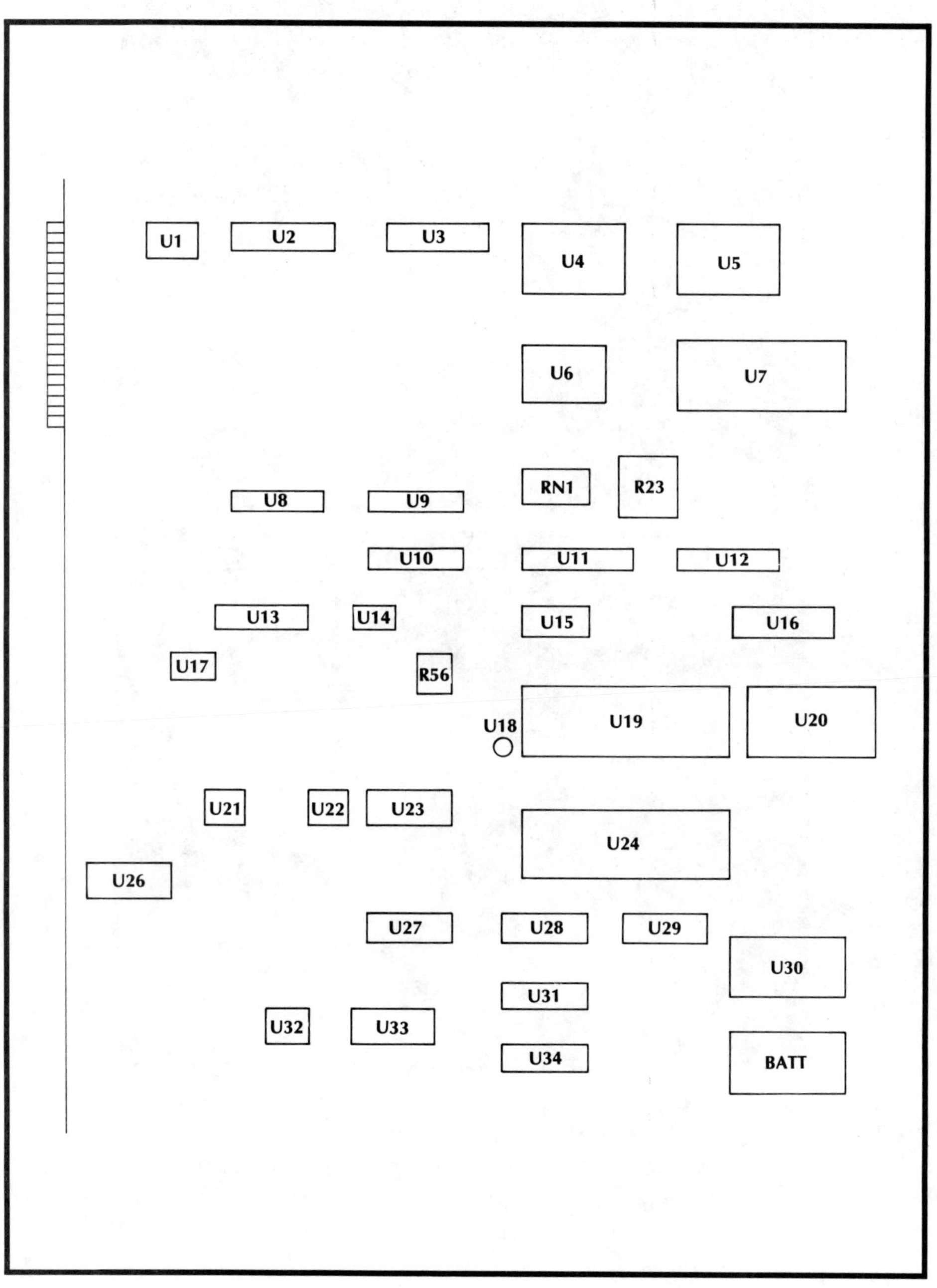
U1
U2
U3
U4
U5
U6
U7
RN1
R23
U8
U9
U10
U11
U12
U13
U14
U15
U16
U17
R56
U18
U19
U20
U21
U22
U23
U24
U26
U27
U28
U29
U30
U31
U32
U33
U34
BATT

Figure 6-30. VideoCipher II Descrambling - Before and After. *The above photo is a scrambled television screen. After processing by the VideoCipher II, the audio and video are in the clear. The black bar across the unscrambled screen is an artifact due to the lack of synchronization between the television field scanning and the 35mm camera shutter.*

Figure 6-29. M/A COM Rack-mounted VideoCipher II-C. *This decoder is designed to be used in commercial headends and only accepts baseband signals. (Courtesy of M/A COM)*

Satellite Receiver Compatibility

There are two types of VideoCipher II units designed for the home market: the model 2000E which has baseband and 70 MHz outputs; and the model 2000 E/B which is exclusively baseband. Only a limited number of 2000E/B receivers were built because the other unit is more versatile. In order to be able to accept a 70 MHz input the 2000E actually has a built-in audio and video receiver so that aside from its lack of tuner, it is essentially a satellite receiver/stereo processor. M/A COM has tested a limited number of home satellite receivers after much controversy suggesting that few were compatible with their decoders. Satisfactory system performance meant the ability to receive authorization codes, to acquire and maintain sync and to provide consumer-acceptable descrambled video and audio signals (at a simulated signal-to-noise input of 8.5 dB). Table 6-3 lists those brands found to be compatible. However, a quote from this fact sheet states:

M/A-COM, Inc.
3033 Science Park Road
San Diego, CA 92121
Tel: (619) 457-2340

RECEIVERS SUCCESSFULLY TESTED WITH VIDEOCIPHER® II SERIES 2000 CONSUMER DESCRAMBLER

Model 2000E - 70 MHz IF Interface

Amplica CSR 100	Chaparral Sierra*	Pico HR 1000*/**
Amplica CSR 200	Drake ESR 24**	Prosat 315 (LNC P-511)
Amplica CSR 300	Drake ESR 224**	ProSat 330*
Anderson Scientific ST 4010	Drake ESR 240**	SatStar Elan
Avcom COM-2A (LNC RDC-11A)	Drake ESR 240A	Sat Tec R-5000
Avcom COM-2B (LNC RDC-11A)	Drake ESR 324	Sigma-Vu Mark IIA
Avcom COM-3 (LNC RDC-3)	Drake ESR 324 Block System*	Sigma-Vu Mark III*
Avcom COM-3R (LNC RDC-3)	Drake ESR 324 S Block System*	Sigma-Vu Mark V*
Birdview MR 20/20***	Drake ESR 424 Block System*	STS MBS-SR
Birdview 20/20 M2***	Drake ESR 524 Block System*	Toki TR220
Birdview 20/20 M3***	Drake Black Widow Series I**	Uniden UST-1000
Birdview 20/20	Drake Black Widow Series II*	Uniden UST-3000
Boman SR-1200	Earth Terminals	Uniden UST-5000*
Boman SR-1500	GCI 8300	Uniden UST-6000*
Boman SR-2500	Houston Tracker System V*	Uniden UST-7000*
Channel Master 6129	Intersat Baby Q	Vector BSR-4000
Channel Master 6130	KLM Skyeye X	Wilson YM400
Channel Master 6131	Lowrance System 70X	Wilson YM450
Channel Master 6134	Luxor 9550	Wilson YM1000
Channel Master 6138	Pico CR 1000*	Winegard RF-1000

Model 2000E/B and 2000E - Composite Video Interface

Arcfinder 2000	General Instrument SRX-207	Panasonic C-1000
Brooks B-250	Janeil BCR-2000***	Panasonic C-2000
Brooks B-250A	Janeil BCR-5000	Panasonic C-2000A
Channel Master 6135	Kenwood KSR-1000***	Panasonic Ku/C-6000
Channel Master 6137	Luxor 9570	Radio Shack SR 2010
Channel Master 6144	Luxor 9900	Radio Shack TDP 900
Drake ESR 424 BLock System	Luxor 9995	Sat Tec R-5100***
Drake ESR 524 Block System	M/A-COM H-1	Scientific Atlanta Homesat 800
Drake ESR-924i	M/A-COM T-1	Standard MT800 - Angile Omni
DX Communications DSB-600	M/A-COM T-2	Uniden UST-5000***
DX Communications DSB-600A	M/A-COM T-6	Uniden UST-6000***
DX Communications DSB-700	NORSAT JR-100	Uniden UST-7000***
DX Communications DSB-700A	NORSAT JR-200	Zenith ZS-3000
DX Communications DSB-700S***	NORSAT XT-100	
DX Communications DSB-800	NORSAT XT-200	

A single sample of each satellite receiver listed above was found to provide acceptable performance and physical/electrical interface to the VideoCipher II Series 2000E or 2000E/B Descrambler when tested under ideal laboratory conditions. Inclusion on this list does not guarantee or warrant perfomrance of all production units of a model under varying environmental conditions. This incomplete list does not imply that receivers not listed have necessarily failed operational testing.

Calll the manufacturer of your receiver if you have any questions.

* Compatible with receiver 70 MHz IF Loop
** Units not tested. Manufacturer certifies no electrical differences between this unit and similar model already tested and approved.
*** Units require factory authorized modification or adjustment for proper operation with VideoCipher II. Contact the manufacturer for additional information.

TABLE 6-4. COMPATIBLE SATELLITE RECEIVERS (Courtesy of M/A COM, Inc.)

Independent testing has indicated that compatibility problems have occasionally arisen, especially when using the VideoCipher 2000-E/B. The descrambler requires an input signal which falls within rather narrow specifications. The composite baseband video output from a satellite receiver must be de-emphasized by the correct amount and the input voltage must fall within the range from 0.5 to 1.5 volts peak-to-peak. Any deviations are not acceptable by the circuitry and are rejected.

Compatibility has obviously not been a concern for those satellite receivers having a built-in VideoCipher II "card" or module such as the M/A COM 2000-R and some of the Channel Master units. These modules are available on an OEM basis to satellite receiver manufacturers.

"A single sample of each satellite receiver ... was found to provide acceptable performance ... when tested under ideal laboratory conditions. Inclusion on this list does not guarantee or warrant performance of all production units of a model under varying environmental conditions. This incomplete list will be periodically updated and does not imply that receivers not listed have necessarily failed operational testing".

REAR PANEL

VHF IN FROM ANT
Connector provided for receiving VHF or local cable service. Type F connector.

VHF OUT TO TV
Connector provided for connecting the descrambler to a television receiver. Type F connector.

CH3/CH4 SWITCH
Selects a VHF output channel from the descrambler on channel 3 or 4. Select the channel that produces the best reception in your area.

70 MHz IN (2000E Only)
Accepts a 70MHz IF satellite signal from a downconverter. Type F connector.

70 MHz OUT (2000E Only)
Loop-through connector that returns the 70 MHz IF Signal from the Descrambler to the satellite receiver. Type F connector.

INPUT
Switches between 70 MHz input and the composite video input.

VIDEO POLARITY
Selects the correct polarity when a 70 MHz IF input is used.

COMPOSITE VIDEO IN
Accepts unclamped, composite video from a satellite receiver. RCA phono connector. (Not required for 70MHz operation.)

CLAMPED VIDEO IN
Accepts clamped, filtered, de-emphasized video from a satellite receiver (Video Out) for non-scrambled satellite broadcasts. RCA phono connector.

COMPOSITE VIDEO INPUT LEVEL
Factory adjustment of video level of satellite broadcast.

CLAMPED VIDEO INPUT LEVEL
Adjusts the brightness level between scrambled and non-scrambled satellite broadcast signals.

VIDEO OUT
Connector provided for connection to a monitor or a VCR. RCA phono connector.

AUDIO IN (R, L)
Accepts audio from a satellite receiver for non-scrambled broadcasts. RCA phono connectors.

LEVEL - AUDIO INPUT
Adjusts the audio level for non-scrambled broadcasts to match scrambled audio levels.

AUDIO OUT (R, L)
Connector provided for connection to a home stereo system or to a stereo-equipped VCR. RCA phono connectors.

STEREO/MONO SWITCH
Switches between mono and stereo inputs.

AUX DATA
For future use.

AUX CLOCK
For future use.

UNIT AUTHORIZATION NUMBER
Unique code that identifies your descrambler.

Courtesy of MA/COM

FRONT PANEL

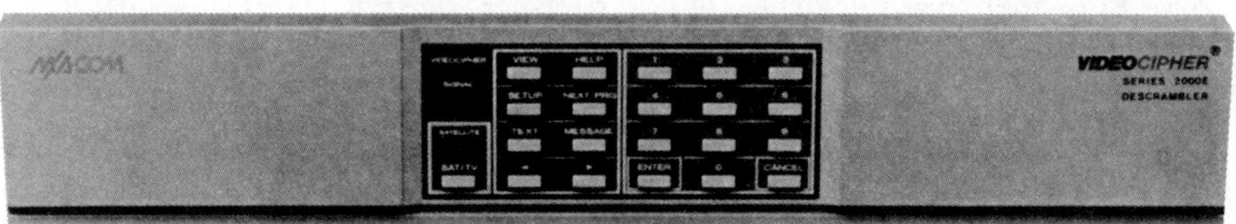

SAT/TV
Switches the descrambler between satellite and local TV programming.

VIEW
Presents the currently scheduled program.

HELP
Provides assistance when using the descrambler features.

SETUP
Used to access stored data and to change certain descrambler operations.

NEXT PRG (Next Program)
Provides information about the next scheduled program.

TEXT
Used to receive broadcast text information from the program supplier including news bulletins, programming promotions and special notices.

MESSAGE
Displays most recently received personal message addressed to your descrambler.

NUMERIC KEYS
Used to enter passwords and to make selections.

ENTER
Used to confirm selections and password entries.

CANCEL
Used to clear any digits that are entered incorrectly.

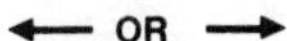

← OR →
Used to select additional screens or to change rating limits.

Courtesy of MA/COM

70 MHz CONNECTION

70MHz Connection (From an LNC or LNA/Downconverter)

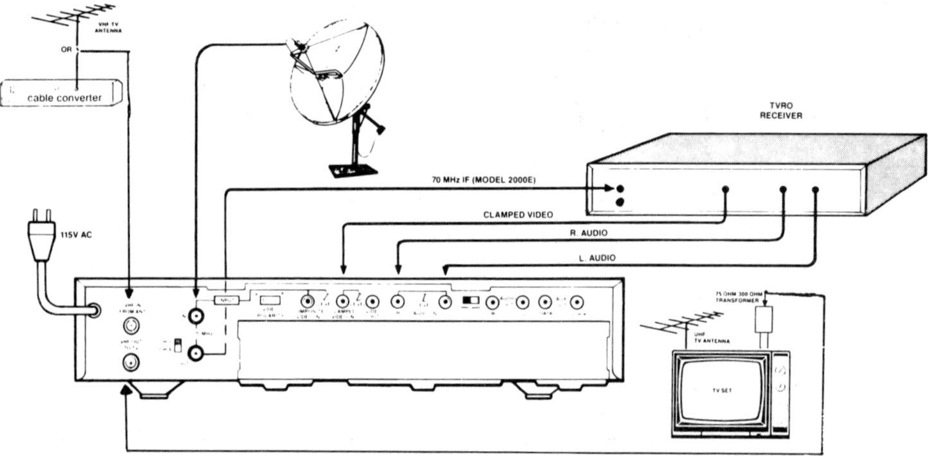

1. Unplug the satellite receiver and the descrambler power supply cords.
2. Disconnect the VHF antenna INPUT and OUTPUT connectors from the satellite receiver. Reconnect these cables to the descrambler connectors VHF IN FROM ANT and VHF OUT TO TV.
3. Disconnect the 70 MHz IF INPUT from the LNC or LNA/downconverter to the satellite receiver. Reconnect this cable to the descrambler connector 70 MHz IN. Position the INPUT switch to the left (when looking at the descrambler rear panel).
4. Connect the accessory cable (Type F male connector each end) to the descrambler connector 70MHz OUT. Connect the remaining cable end to the satellite receiver connector for the 70MHz input (the same receiver connector disconnected in Step 3).
5. Connect the accessory cable (RCA phono connectors each end) to the descrambler connector CLAMPED VIDEO IN and to the corresponding receiver connector. On your receiver this connector may be labeled: VIDEO OUT, TO VCR or TO VIDEO MONITOR.
6. If your receiver is a stereo satellite receiver, connect two accessory cables (RCA male phono connectors each end) to the descrambler connectors AUDIO IN - R (Right) and L (Left). Connect the remaining ends of the accessory cables to the corresponding AUDIO OUT receiver connectors. Then position the AUDIO IN switch to STEREO.

   NOTE: If your receiver is a monaural (MONO) satellite receiver, you may receive stereo audio on scrambled channels with the VIDEOCIPHER descrambler. Connect the receiver's AUDIO OUT connector to either of the descrambler AUDIO IN connectors. Then connect the descrambler to your home stereo system as shown in CONNECTING ACCESSORY COMPONENTS.
7. After completing the wiring procedures, reconnect the descrambler and satellite receiver power supply cords, then refer to TUNING LOCAL PROGRAMS.

Courtesy of MA/COM

DESCRAMBLER INSTALLATION

70 MHz IF Loop Through Connection (From a LNB/Satellite Receiver)

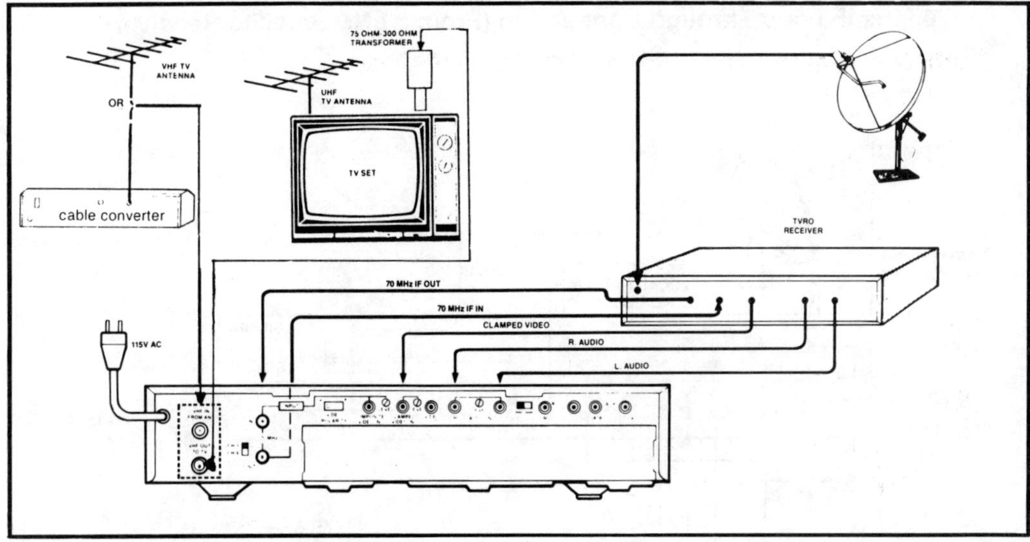

NOTE: Some new LNB-type satellite receivers require a 70 MHz IF loop-through connection. Consult your satellite dealer to determine if your receiver requires this type of installation, then follow the procedure shown on this page. (An additional Type-F male-to-male video accessory cable is required for this type of installation, which can be purchased from your satellite dealer or a local video store).

1. Unplug the satellite receiver and descrambler power supply cords.
2. Connect the cable from your VHF antenna or your local cable service to the descrambler connector, VHF IN FROM ANT. (If you are adding the descrambler to an existing satellite antenna system, disconnect his cable from the receiver and reconnect to the descrambler VHF IN FROM ANT.)
3. Connect a type-F accessory cable to the descrambler connector, VHF OUT TO TV and then to the television's VHF IN connector. (If you are adding the descrambler to an existing satellite antenna system, disconnect this cable from the satellite receiver and reconnect to the descrambler connector VHF OUT TO TV.)
4. Disconnect the existing short jumper cable on the satellite receiver connecting the 70 MHz IF OUT to 70 MHz IF IN.
5. Connect the accessory cable (Type-F male connectors on each end) from the receiver's 70 MHz IF OUT to the descrambler's 70MHz IN connector.
6. Connect an additional accessory cable (Type-F male connector on ach end) from the descrambler's 70 MHz OUT to the satellite receiver's 70 MHz IF IN. Position the INPUT switch to the left (when looking at the descrambler rear panel).
7. Connect the accessory cable (RCA male phono connectors on each end) from the descrambler connector CLAMPED VIDEO IN to the appropriate receiver connector. On your receiver, this connector may be labeled: VIDEO OUT, TO VCR, or TO VIDEO MONITOR.
8. If your satellite receiver has stereo capacity, connect the paired accessory cables (RCA male phono connectors on each end) to the descrambler connectors AUDIO IN - R and L. Connect the remaining ends of accessory cable to the appropriate AUDIO OUT satellite receiver connectors. Then position the descrambler's AUDIO Switch to STEREO.

   If your satellite receiver has monaural (MONO) audio output only, attach one end of the accessory cable to the receiver's AUDIO OUT connector and the remaining cable connector (same color) to either of the descrambler AUDIO IN connectors. Then position the AUDIO switch to MONO.
9. After completing the descrambler installation reconnect the descrambler and satellite receiver power supply cords.

Courtesy of MA/COM

70 MHz IF Loop Through Connection (From a LNB/Satellite Receiver)

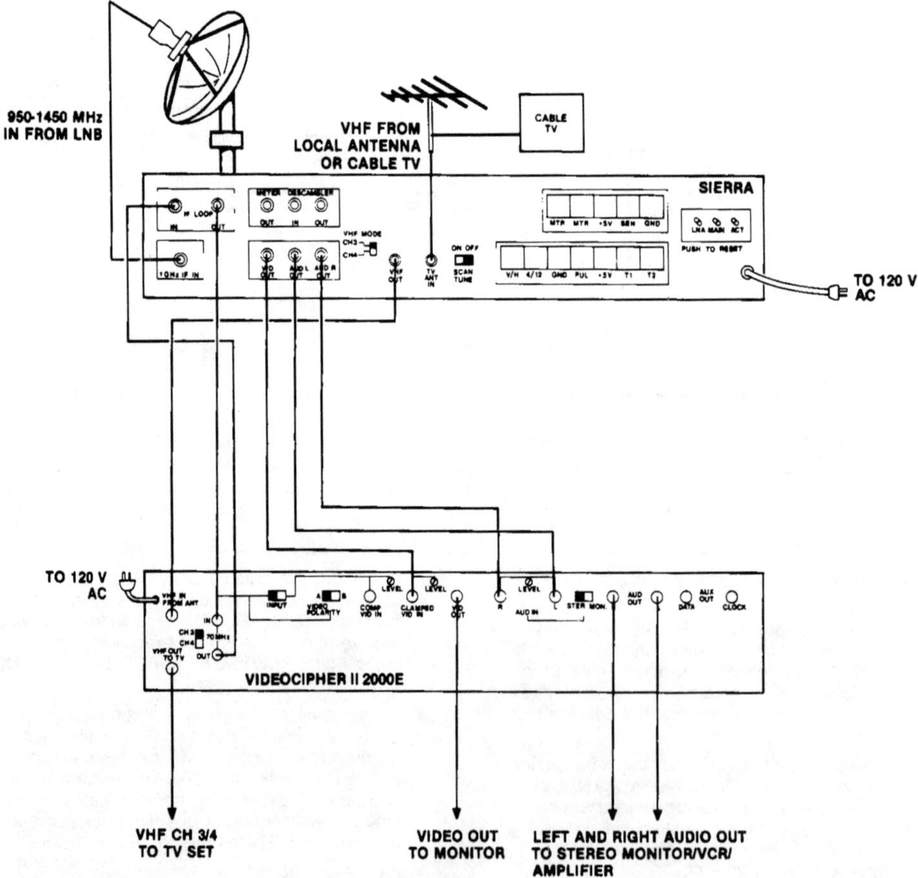

CONNECTION TO A VIDEO MONITOR AND STEREO AMPLIFIER

Refer to the diagram above and connect the VIDEO OUT and AUDIO OUT from the VideoCipher II to the monitor system as shown. Note that the local or cable TV is connected to a separate TV set. This allows viewing satellite TV on the monitor and local or cable TV on the TV simultaneously.

Courtesy of MA/COM

BASEBAND CONNECTION

Baseband Connection (From an LNB/Satellite Receiver)

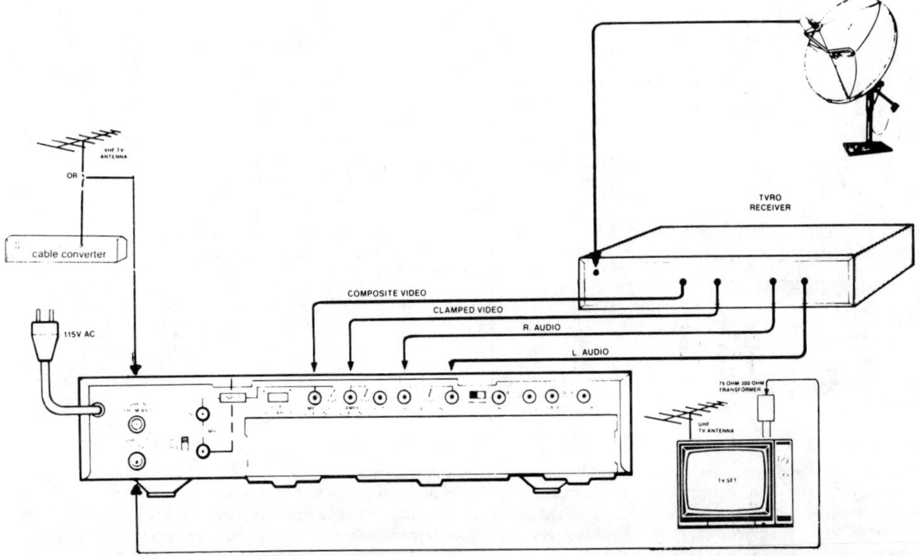

1. Unplug the satellite receiver and descrambler power supply cords.
2. Connect the cable from your VHF antenna or your local cable service to the descrambler connector, VHF IN FROM ANT. (If you are adding the descrambler to an existing satellite antenna system, disconnect this cable from the receiver and reconnect to the descrambler VHF IN FROM ANT.)
3. Connect a Type-F accessory cable to the descrambler connector, VHF OUT TO TV then the television's VHF IN connector. (If you are adding the descrambler to an existing, previously installed antenna system, disconnect this cable from the receiver and reconnect this cable from the receiver and reconnect to the descrambler connector, VHF OUT TO TV.)
4. Connect the RCA-type connector accessory cable from the descrambler connector labeled COMPOSITE VIDEO to the composite video out connector on your receiver. On your satellite receiver this connector may be labeled: TO VIDEOCIPHER, BASEBAND OUT, BB OUT, ACCESSORY OUT, ACCESSORY, COMPOSITE OUT, COMPOSITE VIDEO, VIDEO OUT UNCLAMPED, DECODER OUT, TO DESCRAMBLER or RAW VIDEO.

   Connect the INPUT SWITCH to the right if you have a Series 2000E descrambler.
5. Connect the accessory cable (RCA male phono connectors on each end) from the descrambler connector labeled CLAMPED VIDO IN to the video out connector on your receiver. On your receiver this connector may be labeled: VIDEO OUT, TO VCR, or TO VCR/MONITOR.
6. If your receiver has stereo capacity, connect the paired accessory cables (RCA male phono connectors on each end) to the descrambler connectors AUDIO IN - R (Right) and L (Left) then to the corresponding AUDIO OUT receiver connectors. Then position the AUDIO SWITCH to STEREO.

   If your receiver has monaural (MONO) audio output only, attach one end of the accessory cable to the receiver's AUDIO OUT connector and to either of the descrambler AUDIO IN connectors. Then position the AUDIO SWITCH to MONO.
7. After completing the descrambler installation, reconnect the descrambler and satellite receiver power supply cords.

Courtesy of MA/COM

ACCESSORY COMPONENTS

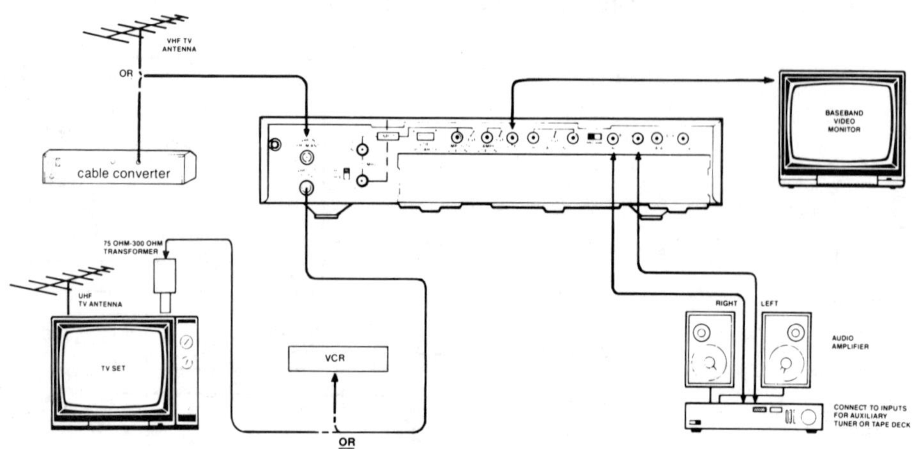

Any accessory equipment used with a standard television and satellite receiver system can also be used with your descrambler. This equipment includes any standard VCR, baseband video monitor and stereo system. The necessary cables to install these items to the descrambler should be supplied with the accessory components; if not, contact your dealer or local electronics supply store. A typical wiring diagram is included below. If you need any additional information, refer to the instructions supplied with your accessory component.

Connecting A VCR

The VCR may be connected to the descrambler VHF OUT to TV and the TV VHF input. If the VCR has baseband audio and video, use the VIDEO and AUDIO OUT connectors. Refer to the VCR instructions for additional information.

Connecting a Cable Converter

The cable converter may be connected to the descrambler VHF IN FROM ANTENNA connector.

Connecting a VHF Antenna (75-Ohm Round Cable)

If your VHF antenna lead-in is a round 75-Ohm cable with a Type-F connector, attach the cable to the descrambler VHF IN connector.

Connecting a VHF antenna (300 Ohm Twin Lead)

If your VHF antenna lead-in is a flat 300 Ohm twin lead, use a 300 to 75 Ohm transformer to connect your VHF antenna to the descrambler VHF IN connector.

Connecting a UHF Antenna

Connect your UHF antenna directly to your television. The descrambler does not affect the antenna operation.

Courtesy of MA/COM

TECHNICAL INFORMATION

The Series 2000E and E/B descramblers use the COMPOSITE VIDEO (unclamped deemphasized) input signal to generate the unscrambled video program. The BASEBAND VIDEO (clamped, filtered, deemphasized) input to the descrambler is used as the normal video receiver output; this signal is recognized by the descrambler as an unscrambled channel, and the unscrambled signal is bypassed directly to the descrambler output. Some receivers do not normally provide the required COMPOSITE VIDEO output and must be modified for use with the VIDEOCIPHER system. Contact your receiver manufacturer or servicing dealer for modification instructions and assistance.

The Model Series 2000E descrambler has an additional input/output 70 MHz IN/OUT, which may be used instead of the COMPOSITE VIDEO input. The Series 2000E descrambler demodulates the 70 MHz IF signal for the descrambler, and also loops back the input to the 70 MHz output for the satellite receiver. This configuration is recommended for all receivers that have a 70 MHz input from a LNC because it requires no modification or adjustment.

Both descramblers provide descrambled baseband video, stereo audio and remodulated VHF outputs, and allow viewing of local VHF signals via the VHF bypass path. Local VHF signals are connected directly to the TV receiver UHF input.

INPUT SECTION

VIDEO FROM SAT RCVR

Signal Format	Negative Sync./Unclamped*
Impedance	75 ohms
Level	0.5 Vpp to 1.5 Vpp

AUDIO FROM SAT RCVR

Impedance	15 kohms min.
Level	1 Vpp

VHF IN FROM ANT

Signal	local VHF
Impedance	75 ohms
	looped through to "VHF OUT TO TV" when "SATELLITE" indicator is off.

OUTPUT SECTION

VIDEO OUT

Signal Format	NTSC, Negative Sync.
Impedance	75 ohms
Level	1 Vpp
Frequency Range	60 Hz to 4.2 MHz
Dispersal at Clamp	40 dB min.

AUDIO OUT

Impedance	1 k ohms max.
Level	2.4 Vpp max.
Frequency Range	30 Hz to 15 kHz

VHF OUT TO TV

Signal	CH 3 or CH 4 changeable or "VHF IN FROM ANT" input
Impedance	75 ohms

BASEBAND OUT

Signal Format and Level	Same as "VIDEO FROM SAT RCVR"
Impedance	75 ohms

DATA/CLOCK

Signal Format	Serial digital data at 88 kbps and an associated clock
Level	TTL

PRIMARY POWER

Source	115 C ± 10%, 60 Hz ± 5%
Power Consumption	40 watts

MECHANICAL

Size	17.12" (W) x 10.5" (D) + 2.95" (H)
Weight	20 lbs.

OTHER FEATURES

Video gain adjustment	Rear Panel Pot.

All specifications are subject to change without notice.

*Consult your satellite receiver instruction manual to determine if an unclamped output is available. If not, contact the manufacturer for modification instructions.

Courtesy of MA/COM

F. SCIENTIFIC ATLANTA B-MAC

The B-MAC system is used as an example of the multiplexed analog component transmissions which were explored in Chapter IV. This system is being implemented in Australia and is being considered in other countries including the United States.

In B-MAC, the 63.5 microsecond scan line is composed of 11.0 microseconds of data, 35 microseconds of luminance signal and 17.5 microseconds of time-compressed chrominance signal. By comparison, the luminance signal occupies 52.5 microseconds on a standard NTSC line. A total of 1.86 million bits of data are transmitted per second with the video signal, providing digital audio, control and additional data services.

Use of the Horizontal Blanking Interval

1.573 Mbits/second are relayed by the horizontal blanking interval. Six channels of audio and associated error detection and correction bits use 1.510 Mbits/second with a remainder of 62.9 Mbits/second allocated to a permanent business data channel. A separate 9600 bits/second RS232 port is available for connection to a standard microcomputer. Four of the six 204 kilobits/second audio channels can be optionally reassigned to data delivery. In fact, the entire video interval can be transformed into a 10.8 Mbits/second business data trunk line. The allocation of these resources are entirely under the control of the system operator through the flexible addressability feature.

Six channels of high fidelity audio can be transmitted over each channel. These have frequency responses from 20 hertz to 18 KHz where the output is reduced in level by 3 decibels.

Use of the Vertical Blanking Interval

All system control data is transmitted in the vertical blanking interval. Lines 1 through 8 relay bits controlling clock and sync recovery as well as subscriber addresses. Lines 9 through 13 carry text relayed in each line by a single row of 40 ASCII characters. The ASCII convention is an eight bit code yielding a possible combination of 256 charac-

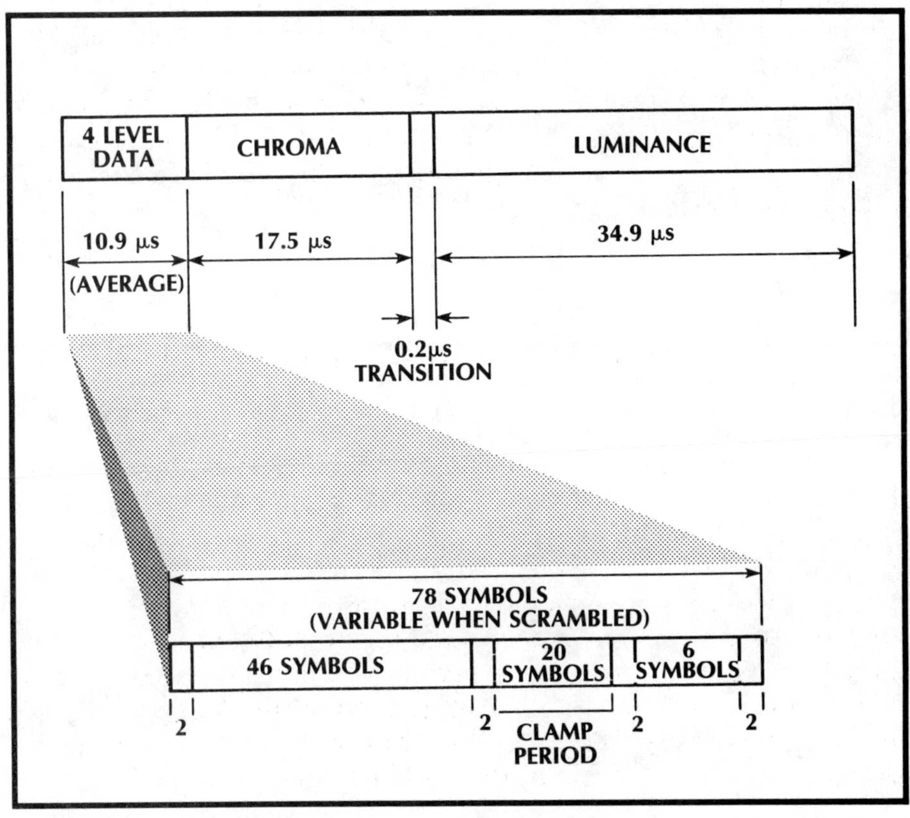

Figure 6-31. Structure of the B-MAC Signal. *Each line transmitted via the B-MAC format is structured as indicated here. In an unscrambled line there are 50 useful data symbols. However, in a scrambled line the number of useful symbols may vary from 42 to 58 depending upon the amount of line shifting and the color burst may be positioned pseudo-randomly among the useful data symbols.*

ters. It is the standard language used by microcomputers. The text can be molded into a variety of forms. These include 200 pages of standard text with a 20 second access time, film subtitles or closed captions for the hearing impaired, lists of pay-per-view specials or forced messages which appear when necessary on the screen in times of emergency. Messages can be addressed to an individual, a state, a city or any region providing a true "electronic mail" capacity.

Figure 6-32. Scientific Atlanta B-MAC Series 9700 Decoder. *This rack mounted decoder is used in conjunction with a commercial grade satellite receiver. (Courtesy of Scientific Atlanta).*

Figure 6-33. Scientific Atlanta B-MAC Model 9730 Integrated Receiver/Decoder. *This unit accomplishes the tasks decoding and receiving B-MAC satellite TV broadcasts. (Courtesy of Scientific Atlanta)*

TABLE 6-5. CONTROL DATA LOCATION ON B-MAC
Vertical Blanking Interval

LINE	DESIGNATION	DESCRIPTION
1	Clock Recovery	Resets the system sync clock
2	Sync Recovery	Codes the beginning of the sync
3	System Data	Defines the services being relayed
4-8	User Address Data	Of 377 bits, 28 for addressing up to 268 million subscribers
9-13	Teletext	Used to relay teletext
14-16	Teletext or Spare	Used to relay teletext or as a spare
17-20	Operational Test Signals or Teletext	Checks whether link and decoder are operating and is optionally used for teletext
21	Quiet line	Sets gray reference level
22	Video AGC Reference	Sets reference level for automatic gain control

System Uplink Operation

Four uplink operational levels, 0,1,2, and 3 are defined, each of which has a set of hardware requirements and capabilites. The features of level 3, controlled by a Hewlett Packard 1000 and an EMC/2 computer, are outlined here as an instructive example. This level has among others, the following features:

- Automatic addressing of decoders
- Individual or group control of decoder services
- Individual or group control of personal messages
- Independent support for three broadcasters

- Central monitoring and fault reporting
- Diagnostic logging to disk files for historical record
- System data logging
- Operator interface with supervisory control of subscriber authorization functions
- Automatic broadcast control interface
- Controlled access to video, audio, teletext, utility data and auxiliary data to maximum 7 million receive sites
- Business system interface at maximum transaction rate of 60 per second
- Maximum addressing rate of 12,540 per second

TABLE 6-6. SOME USERS OF B-MAC FOR HI-NET NETWORK (Ku-Band Satellite)

USER	SATELLITE/TRANSPONDER
ESPN	G-STAR 1/20
Showtime West	G-STAR 1/19
WTBS	G-STAR 1/32
CNN Headline	G-STAR 1/31

APPENDIX A: WORLDWIDE BROADCAST FORMATS

Satellite television links nations of the world more closely together than ever before. However, in the process, the differences in translating between the three principal broadcast formats becomes an important factor. The table below outlines the major features of PAL, SECAM and NTSC as well as the format and standard for both VHF and UHF broadcasts and the line voltage and frequency in use in countries around the world. Those cases where a standard for satellite service has been established are indicated. This table should be used in conjunction with those presented in Appendix B.

WORLDWIDE TELEVISION BROADCAST STANDARDS

COUNTRY	BROADCASTING STANDARD VHF	BROADCASTING STANDARD UHF	COLOR SYSTEM	SATELLITE SERVICE STANDARD	LINE VOLTAGE/ FREQUENCY
Algeria	B		PAL	PAL	127/220/380-50
Argentina	N	N	PAL	PAL	200-50
Australia	B		PAL	PAL	240/415-50
Austria	B	G	PAL		220/380-50
Barbados	M		NTSC		110/220-50
Belgium	B	H	PAL		220-50
Bermuda	M		NTSC		115/230-60
Bolivia	N		NTSC		110/220-50
Brazil	M	M	PAL	PAL	110/127/220-50/60
Bulgaria	D	K	SECAM	SECAM	220-50
Canada	M	M	NTSC	NTSC	115/230-60
Canary Islands	B		PAL		110/220-50
Chile	M	M	NTSC		220/380-50
China	D	K	PAL		220-50
Colombia	M		NTSC	NTSC	110/220-60
Congo	D				220-50
Costa Rica	M		NTSC		120/220-60
Cuba	M	M	NTSC	NTSC	115/220-60
Cyprus	B	H	PAL		240-50
Czechoslovakia	D	K	SECAM	SECAM	220-50
Denmark	B	G	PAL		220/380-50
Dominican Rep.	M	M	NTSC		110-60
East Germany	B	G	SECAM	SECAM	220/380-50
Ecuador	M		NTSC		110/220-60
Egypt	B		SECAM		220-50
El Salvador	M		NTSC		110/220-60
England	A	I	PAL	PAL	240/415-50
Ethiopia	B		PAL		127/220-50
Finland	B	G	PAL		220/380-50
France	E	L	SECAM	SECAM	various-50
Gabon	K		SECAM		220-50
Ghana	G		PAL		230-50
Gibraltar	B		PAL		240-50
Greece	B		SECAM		220-50
Greenland	M		NTSC		220/380-50
Guadeloupe	K		SECAM		220-50
Guam	M		NTSC		110/220-60

Guatemala	M		NTSC		110/220/127/220-60
Guyana	M		NTSC		220-50
Haiti	M		SECAM		110/220-50/60
Honduras	M		NTSC		110-60
Hong Kong		I	PAL		200-50
Hungary	D	K	SECAM	SECAM	220-50
Iceland	B	G	PAL		220-50
India	B		PAL	PAL	230/400-50
Indonesia	B		PAL	PAL	127/220-50
Iran	B	G	SECAM		220-50
Iraq	B		SECAM		220/380-50
Ireland	I/A		PAL		220/380-50
Israel	B	G	PAL		230/400-50
Italy	B	G	PAL	PAL	220/380-50
Ivory Coast	K		SECAM		220/380-50
Jamaica	M				110/220-50
Japan	M	M	NTSC	NTSC	110/220-50/60
Jordan	B		PAL		220-50
Kenya	B		PAL		240-50
Korea (South)	M	M	NTSC		110/60
Korea (North)	D		PAL		220-50
Kuwait	B		PAL		240-50
Lebanon	B		SECAM		110/190-50
Liberia	B		PAL		120-60
Libya	B		SECAM		127/130-50
Luxembourg	C/G	L	SECAM		110/220-50
Malawi	B				230-50
Malaysia	B		PAL	PAL	230/400-50
Malta	B		PAL		240-50
Mexico	M		NTSC	NTSC	110/220-60
Micronesia	M		NTSC		
Monaco	E	G	PAL/SECAM		127/220-50
Mongolia	D				220-50
Morocco	B		SECAM	SECAM	127/220-50
Netherlands	B	G	PAL	PAL	220-50
New Caledonia	K		SECAM		220-50
New Zealand	B		PAL		230/400-50
Nicaragua	M		NTSC		120-60
Niger	K		SECAM	SECAM	220/380-50
Nigeria	B		PAL	PAL	230/400-50
Norway	B	G	PAL	PAL	230-50
Oman	B	G	PAL	PAL	220-50
Pakistan	B		PAL		230/400-50
Panama	M		NTSC		110/115/120/126-60

Paraguay	N		NTSC		220-50
Peru	M		NTSC	NTSC	220-60
Philippines	M	M	NTSC	NTSC	110/220-60
Poland	D	K	SECAM	SECAM	220/380-50
Portugal	B	G	PAL	PAL	220/380-50
Puerto Rico	M	M	NTSC		120-60
Qatar	B		PAL		220-50
Romania	D		SECAM		220-50
Rwanda	K				220-50
Samoa-American	M		NTSC		230-50
Saudi Arabia	B		PAL/SECAM	SECAM	127/220/380-50/60
Senegal	B		SECAM		127/220-50
Sierra Leone	B		PAL		230-50
Singapore	B	G	PAL		230/400-50
South Africa	I	I	PAL	PAL	220/380-50
Spain	B	G	PAL	PAL	127/220/380-50
Sri Lanka	B		PAL		220-50
Sudan	B			PAL	240-50
Surinam	M		NTSC		110/115/127/220-60
Sweden	B	G	PAL		220/380-50
Switzerland	B	G	PAL		220/380-50
Syria	B		SECAM		115/200-50
Tahiti	K		SECAM		127-60
Taiwan	M		NTSC		110/200-60
Tanzania	B		PAL		230/440-50
Thailand	B	G	PAL	PAL	230/380-50
Trinidad/Tobago	M		NTSC		115/230-60
Tunisia	B		SECAM		115/220/380-50
Turkey	B	G	PAL		220/380-50
U.S.A.	M	N	NTSC	NTSC	110/220-60
U.S.S.R.	D	K	SECAM	SECAM	127/220-50
Uganda	B				240/415-50
United Arab Emirates	B	G	PAL		220-50
Upper Volta	K				220/380-50
Uruguay	N		PAL		220-50
Venezuela	M		NTSC	NTSC	120/240-60
Virgin Islands	M		NTSC		110-60
West Germany	B	G	PAL	PAL	220/380-50
Yemen	B				230-50
Yugoslavia	B	G	PAL		220/380-50
Zaire	K		SECAM	SECAM	220-50
Zambia	B		PAL		220-50
Zimbabwe	B		PAL		230/240-50

KEY TO TABLE:

BROADCASTING SYSTEM	NUMBER OF LINES	CHANNEL BANDWIDTH (MHz)	AUDIO TO VIDEO CARRIER SEPARATION (MHz)
A	405	5	-3.5
B	625	7	5.5
C	625	7	5.5
D	625	8	6.5
E	819	14	11.15
F	819	7	5.5
G,H	625	8	5.5
I	625	8	6.0
K	625	8	6.5
L	625	8	6.5
M	525	6	4.5
N	525	6	4.5

APPENDIX B: WORLDWIDE TELEVISION CHANNEL STRUCTURES

This table details channel designation and bandwidth as well as the center frequencies of both the audio and video carrier for worldwide television systems.

I. VHF CHANNEL STRUCTURES

BAND	CHANNEL NUMBER	BAND LIMITS (MHz)	VIDEO CARRIER FREQUENCY	AUDIO CARRIER FREQUENCY
		Standard A - Great Britain (5 MHz)		
	B 1	41.25 - 46.25	45.00	41.50
	B 2	48 - 53	51.75	48.25
I	B 3	53 - 58	56.75	53.25
	B 4	58 - 63	61.75	58.25
	B 5	63 - 68	66.75	63.25
	B 6	176 - 181	179.75	176.25
	B 7	181 - 186	184.75	181.25
	B 8	186 - 191	189.75	186.25
III	B 9	191 - 196	194.75	191.25
	B 10	196 - 201	199.75	196.25
	B 11	201 - 206	204.75	201.25
	B 12	206 - 211	209.75	206.25
	B 13	211 - 216	214.75	211.25
	B 14	216 - 221	219.75	216.25
		Standard B - Australia (7 MHz)		
IF	—	33.15 - 40.15	38.90(a)	33.40(a)
	0	45 - 52	46.25	51.75
I	1	56 - 63	57.25	62.75
	2	63 - 70	64.25	69.75
	3	85 - 92	86.25	91.75
II	4	94 - 101	95.25	100.75
	5	101 - 108	102.25	107.75
	5A	137 - 144	138.25	143.25
	6	174 - 181	175.25	180.75
	7	181 - 188	182.25	187.75

III	8	188 - 195	189.25	194.75
	9	195 - 202	196.25	201.75
	10	208 - 215	209.25	214.75
	11	215 - 222	216.25	221.75

Standard B,C - Europe (7 MHz)

IF	—	33.15 - 40.15	38.90	33.40
	E 2	47 - 54	48.25	53.75
I	E 3	54 - 61	55.25	60.75
	E 4	61 - 68	62.25	67.75
	E 5	174 - 181	175.25	180.75
	E 6	181 - 188	182.25	187.75
	E 7	188 - 195	189.25	194.75
III	E 8	195 - 202	196.25	201.75
	E 9	202 - 209	203.25	208.75
	E 10	209 - 216	210.25	215.75
	E 11	216 - 223	217.25	222.75
	E 12	223 - 230	224.25	229.75

Standard B,C - Special cable TV channels Europe (7 MHz)

IF	—	33.15 - 40.15	38.90	33.40
	S 1	104 - 111	105.25	110.75
	S 2	111 - 118	112.25	117.75
	S 3	118 - 125	119.25	124.75
	S 4	125 - 132	126.25	131.75
L(b)	S 5	132 - 139	133.25	138.75
	S 6	139 - 146	140.25	145.75
	S 7	146 - 153	147.25	152.75
	S 8	153 - 160	154.25	159.75
	S 9	160 - 167	161.25	166.75
	S 10	167 - 174	168.25	173.75
	S 11	230 - 237	231.75	236.75
	S 12	237 - 244	238.25	243.75
	S 13	244 - 251	245.25	250.75
	S 14	251 - 258	252.25	257.75

U(b)	S 15	258 - 265	259.25	264.75
	S 16	265 - 272	266.25	271.75
	S 17	272 - 279	273.25	278.75
	S 18	279 - 286	280.25	285.75
	S 19	286 - 293	287.25	292.75
	S 20	293 - 300	294.25	299.75

Standard B - Italy (7 MHz)

IF	—	31.15 - 40.15	38.90	33.40
I	A	52.5 - 59.5	53.75	59.25
	B	61 - 68	62.25	67.75
II	C	81 - 88	82.25	87.75
	D	174 - 181	175.25	180.75
	E	182.5 - 189.5	183.75	189.25
	F	191 - 198	192.25	197.75
III	G	200 - 207	201.25	206.75
	H	209 - 216	210.25	215.75
	H1	216 - 223	217.25	222.75
	H2	223 - 230	224.25	229.75

Standard B - Morocco (7 MHz)

IF	—	33.15 - 40.15	38.90	33.40
	M 4	162 - 169	163.25	168.75
	M 5	170 - 177	171.25	176.75
	M 6	178 - 185	179.25	184.75
III	M 7	186 - 193	187.25	192.75
	M 8	194 - 201	195.25	200.75
	M 9	202 - 209	203.25	208.75
	M 10	210 - 217	211.25	216.75

Standard B - New Zealand (7 MHz)

IF	—	33.15 - 40.15	38.90	33.40
	1	44 - 51	45.25	50.75

I	2	54 - 61	55.22	60.75
	3	61 - 68	62.25	67.75
	4	174 - 181	175.25	180.75
	5	181 - 188	182.25	187.75
	6	188 - 195	189.25	194.75
III	7	195 - 202	196.25	201.75
	8	202 - 209	203.25	208.75
	9	209 - 216	210.25	215.75
	10	216 - 223	217.25	222.75
		Standard D - China (8 MHz)		
IF	—	30.25 - 38.25	37.00	30.50
	1	48.5 - 56.5	49.75	56.25
	2	56.5 - 64.5	57.75	64.25
I	3	64.5 - 72.5	65.75	72.25
	4	76.0 - 84.0	77.25	83.75
	5	84.0 - 92.0	85.25	91.75
	6	167 - 175	168.25	174.75
	7	175 - 183	176.25	182.75
	8	183 - 191	184.25	190.75
III	9	191 - 199	192.25	198.75
	10	199 - 207	200.25	206.75
	11	207 - 215	208.25	214.75
	12	215 - 223	216.25	222.75
		Standard D - OIRT (8 MHz)		
IF(c)	—	32.15 - 40.15	38.90	32.40
	R I	48.5 - 56.5	49.75	56.25
I	R II	58 - 66	59.25	66.75
	R III	76 - 84	77.25	77.25
II	R IV	84 - 92	85.25	85.25
	R V	92 - 100	93.25	93.25

	R VI	174 - 182	175.25	175.25
	R VII	182 - 190	183.25	183.25
	R VIII	190 - 198	191.25	191.25
III	R IX	198 - 206	199.25	199.25
	R X	206 - 213	207.25	207.25
	R XI	214 - 222	215.25	215.25
	R XII	222 - 230	223.25	223.25
		Standard E - France (13.15 MHz)		
IF	—	26.30 - 39.45	28.05	39.20
I	F 2	41.00 - 54.15	52.40	41.25
	F 4	54.15 - 67.30	65.55	54.40
	F 5	162.25 - 175.15	164.00	175.15
	F 6	162.00 - 175.15	173.40	162.25
	F 7	175.40 - 188.55	177.15	188.30
III	F 8A	174.00 - 188.00	185.25	174.10
	F 8	175.15 - 188.30	186.55	175.40
	F 9	188.55 - 201.70	190.30	201.45
	F 10	188.30 - 201.45	199.70	188.55
	F 11	201.70 - 214.85	203.45	214.60
	F 12	201.45 - 214.60	212.85	201.70
		Standard I - Ireland (8 MHz)		
IF	—	21.15 - 40.15	38.90(d)	32.90(e)
	1 A	44.5 - 52.5	45.75	51.75
I	1 B	52.5 - 60.5	53.75	59.75
	1 C	60.5 - 68.5	61.75	67.75
	1 D	174 - 182	175.25	181.25
	1 E	182 - 190	183.25	189.25
III	1 F	190 - 198	191.25	197.75
	1 G	198 - 206	199.25	205.25
	1 H	206 - 214	207.25	213.25
	1 J	214 - 222	215.25	221.25

		Standard I - South Africa (8 MHz)		
IF	—	32.15 - 40.15	38.90	32.90
	4	174 - 182	175.25	181.25
	5	182 - 190	183.25	189.25
	6	190 - 198	191.25	197.25
	7	198 - 206	199.25	205.25
III	8	206 - 214	207.25	213.25
	9	214 - 222	215.25	221.25
	10	222 - 230	223.25	229.25
	11	230 - 238	231.25	237.25
	12	238 - 246	—not in use—	—not in use—
	13	246 - 254	247.43	253.43
		Standard K1 - French OPTA (8 MHz)		
IF	—	31.45 - 39.45	32.70	39.20
	4	174 - 182	175.25	181.75
	5	182 - 190	183.25	189.75
III	6	190 - 198	191.25	197.75
	7	198 - 206	199.25	205.75
	8	206 - 214	207.25	213.75
	9	214 - 222	215.25	221.75
		Standard L - France (8 MHz)		
IF	—	31.45 - 39.45	32.70	39.20
	A	41 - 49	47.75	41.25
I	B	49 - 57	55.75	49.25
	C	57 - 65	63.75	57.25
	C1	53.75 - 61.75	60.50	54.00
	1	174.75 - 182.75	176.00	182.50
	2	182.75 - 190.75	184.00	190.50
III	3	190.75 - 198.75	192.00	198.50
	4	198.75 - 206.75	200.00	206.50
	5	206.75 - 214.75	208.00	214.50
	6	214.75 - 222.75	216.00	222.50

		Standard M - Japan (6 MHz)		
IF	—	41.00-47.00	45.75	41.25
	J1	90-96	91.25	95.75
II	J2	96-102	97.25	101.75
	J3	102-108	103.25	107.75
	J4	170-176	171.25	175.75
	J5	176-182	177.25	181.75
	J6	182-188	183.25	187.75
	J7(g)	188-194	189.25	193.75
III	J8(h)	192-198	193.25	197.75
	J9	198-204	199.25	203.75
	J10	204-210	205.25	209.75
	J11	210-216	211.25	215.75
	J12	216-222	217.25	221.75
		Standard M,N - USA (6 MHz)		
IF	—	41.00-47.00	45.75	41.25
	A02	54-60	55.25	59.75
	A03	60-66	61.25	65.75
I	A04	66-72	67.25	71.75
	A05	76-82	77.25	81.75
	A06	82-88	83.25	87.75
	A07	174-180	175.25	179.75
	A08	180-186	181.25	185.75
	A09	186-192	187.25	191.75
III	A10	192-198	193.25	197.75
	A11	198-204	199.25	203.75
	A12	204-210	205.25	209.75
	A13	210-216	211.25	215.75

(a) also 36.875 and 31.375 MHz, respectively
(b) L - lower and U - upper channels
(c) USSR - 31.25 to 39.25/38.0/31.5 MHz
(d) Great Britain also uses 39.5 and 33.5 MHz, respectively
(e) Channel spacing of 4 MHz II.

UHF CHANNEL STRUCTURES

BAND	CHANNEL NUMBER	BAND LIMITS (MHz)	VIDEO CARRIER FREQUENCY	AUDIO CARRIER FREQUENCY
		Standard M - Japan (6 MHz)		
IF	—	41.0-47.0	45.75	41.25
	45	662-668	663.25	667.75
	46	668-674	669.25	573.75
	47	674-680	675.25	679.75
	48	680-686	681.25	685.75
	49	686-692	687.25	691.75
	50	692-698	693.25	697.75
	51	698-704	699.25	703.75
	52	704-710	705.25	709.75
V	53	710-716	711.25	715.75
	54	716-722	717.25	721.75
	55	722-728	723.25	727.75
	56	728-734	729.25	733.75
	57	734-740	735.25	739.75
	58	740-746	741.25	745.75
	59	746-752	747.25	751.75
	60	752-758	753.25	757.75
	61	758-764	759.25	763.75
	62	764-770	765.25	769.75
		Standard K - China (8 MHz)		
IF	—	30.25-38.25	37.00	30.50
	13	470-478	471.25	477.75
	14	478-486	479.25	485.75
	15	486-494	487.25	493.75
	16	494-502	495.25	501.75
	17	502-510	503.25	509.75
IV	18	510-518	511.25	517.75
	19	518-526	519.25	525.75
	20	526-534	527.25	533.75
	21	534-542	535.25	541.75

	22	542-550	543.25	549.75
	23	550-558	551.25	557.75
	24	558-566	559.25	565.75
		Free for radio astronomy		
	25	604-612	605.25	611.75
	26	612-620	613.25	619.75
	27	620-628	621.25	627.75
	28	628-636	629.25	635.75
V	34	676-684	677.25	683.75
	35	684-692	685.25	691.75
	36	692-700	693.25	699.75

Standard M,N - Americas (6 MHz)

IF	—	41.0-47.0	45.75	41.25
	14	470-476	471.25	475.75
	15	476-482	477.25	481.75
IV				
	41	632-638	633.25	637.75
	42	638-644	639.25	643.75
	43	644-650	645.25	649.75
	44	650-656	651.25	655.75
	45	656-662	657.25	661.75
	46	662-668	663.25	667.75
	47	668-674	669.25	673.75
V		like Japan's channels 47 to 60		
	62	758-764	759.25	763.75
	63	764-770	765.25	769.75
	64	770-776	771.25	775.75
	65	776-782	777.25	781.75
	66	782-788	783.25	787.75
	67	788-794	789.25	793.75
	68	794-800	795.25	799.75
	69	802-806	801.25	805.75

BAND	CHANNEL NUMBER	BAND LIMITS (MHz)	VIDEO CARRIER FREQUENCY	AUDIO G,H	CARRIER I	FREQUENCY K,L
		Standards G,H,I,K,L — CCIR Standard (8 MHz)				
		IF - same as VHF IF in each corresponding country				
	21	470-478	471.25	476.75	477.25	477.75
	22	478-486	479.25	484.75	485.25	485.75
	23	486-494	487.25	492.75	493.25	493.75
	24	494-502	495.25	500.75	501.25	501.75
	25	502-510	503.25	508.75	509.25	509.75
	26	510-518	511.25	516.75	517.25	517.75
	27	518-526	519.25	524.75	525.25	525.75
	28	526-534	527.25	532.75	533.25	533.75
IV	29	534-542	535.25	540.75	541.25	541.75
	30	542-550	543.25	548.75	549.25	549.75
	31	550-558	551.25	556.75	557.25	557.75
	32	558-566	559.25	564.75	565.25	565.75
	33	566-574	567.25	572.75	573.25	573.75
	34	574-582	575.25	580.75	581.25	581.75
	35	582-590	583.25	588.75	589.25	589.75
	36	590-598	591.25	596.75	597.25	597.75
	37	598-606	599.25	604.75	605.25	605.75
	38	606-614	607.25	612.75	613.25	613.75
	39	614-622	615.25	620.75	621.25	621.75
	40	622-630	623.25	628.75	629.25	629.75
	41	630-638	631.25	636.75	637.25	637.75
	42	638-646	639.25	644.75	645.25	645.75
	43	646-654	647.25	652.75	653.25	653.75
	44	654-662	655.25	660.75	661.25	661.75
	45	662-670	663.25	668.75	669.25	669.75
	46	670-678	671.25	676.75	677.25	677.75
	47	678-686	679.25	684.75	685.25	685.75
	48	686-694	687.25	692.75	693.25	693.75
	49	694-702	695.25	700.75	701.25	701.75
	50	702-710	703.25	708.75	709.25	709.75

V	51	710-718	711.25	716.75	717.25	717.75
	52	718-726	719.25	724.75	725.25	725.75
	53	726-734	727.25	732.75	733.25	733.75
	54	734-742	735.25	740.75	741.25	741.75
	55	742-750	743.25	748.75	749.25	749.75
	56	750-758	751.25	756.75	757.25	757.75
	57	758-766	759.25	764.75	765.25	765.75
	58	766-774	767.25	772.75	773.25	773.75
	59	774-782	775.25	780.75	781.25	781.75
	60	782-790	783.25	788.75	789.25	789.75
	61	790-798	791.25	796.75	797.25	797.75
	67	838-846	839.25	844.75	845.25	845.75
	68	846-854	847.25	852.75	853.25	853.75

APPENDIX C. THE DECIBEL NOTATION

Decibels express the relative values of two signals. The logarithmic scale is used to compress large differences in numbers to a more manageable range. Decibels are defined by the following equation:

Decibel difference = 10 logarithm (signal A divided by signal B)

For example, if signal A is 1000 watts and signal B is 10 watts, then signal A is 20 dB stronger than signal B because:

$$\begin{aligned}\text{Decibel difference} &= 10 \log (1000/10) \\ &= 10 \times 2 \\ &= 20\end{aligned}$$

Therefore, if an amplifier received a signal of 10 watts and increased its power to 1000 watts, it would have a gain of 20 decibels. Similarly, if a 10 watt signal was increased by a factor of 1,000,000 to 10 million, the gain would be 60 dB.

Decibels are also expressed relative to a reference value such as watts, milliwatts or millivolts. The abbreviations dBw, dBm and dBmv mean the relative increase relative to one watt, one milliwatt or one millivolt, respectively. For example, 20 dBmv means a power of 100 millivolts while 60 dBw means a power of 1 million watts.

APPENDIX D. GLOSSARY OF TERMS

Absolute Zero

The coldest possible temperature at which all molecular motion ceases. It is expressed in degrees kelvin as measured from absolute zero. Zero °K equals minus 273.16 °C or minus 459.69°F.

Alignment

The process of fine tuning a dish or an electronic circuit to maximize its sensitivity and signal receiving capability.

Amplifier

A device used to increase the power of a signal.

Analog-to-Digital Converter

A circuit that converts analog signals to an equivalent digital form. The varying analog signal is sampled at a series of points in time. The voltage at each of these points is then represented by a series of digits. The higher this sampling frequency, the finer are the gradations and the more accurately is the signal represented.

Antenna

A device that collects and focuses electromagnetic energy, i.e., contributes an energy gain. Gain is proportional to surface area for a microwave dish.

Antenna Efficiency
The percentage of incoming satellite signal actually captured by an antenna.

Antenna Illumination
Describes how a feedhorn "sees" the surface of a dish as well as the surrounding terrain.

Aperture
The collection area of a parabolic antenna.

Aspect Ratio
The ratio of television screen width to height. The standard aspect ratio is 4 to 3.

Attenuator
A passive device which reduces the power of a signal.

Audio Subcarrier
The carrier wave that transmits audio information on a broadcast signal. Satellite transmissions relay audio subcarriers in the frequency range between 5 and 8.5 MHz.

Automatic Brightness Control
A television circuit used to automatically adjust picture tube brightness in response to changes in background or ambient light.

Automatic Fine Tuning
A circuit that automatically maintains the correct tuner oscillator frequency and compensates for drift and for moderate amounts of mistuning.

Automatic Frequency Control (AFC)
A circuit that locks onto a chosen frequency and will not drift away from that frequency.

Automatic Gain Control (AGC)
A circuit that locks the gain onto a fixed value and thus compensates for varying input signal levels keeping the output constant.

Azimuth-Elevation (Az-El) Mount
An antenna mount which tracks satellites by moving in two directions: the azimuth in the horizontal plane; and elevation up from the horizon.

Azimuth
Degrees of rotation clockwise from true north.

Back Porch
That portion of the horizontal blanking pulse that follows the trailing edge of the horizontal sync pulse.

Bandpass Filter
A circuit or device that allows only a specified range of frequencies to pass from input to output.

Bandwidth
The frequency range allocated to any communication circuit.

Beamwidth
A measure used to describe the width of vision of an antenna. It is measured as degrees between the 3 dB half power points.

Blanking Pulse Level
The reference level for video signals. The blanking pulses must be aligned at the input to the picture tube.

Blanking Signal
Pulses used to extinguish the scan illumination during horizontal and vertical retrace periods.

Block Downconversion
The process of lowering the entire band of frequencies in one step to some intermediate range to be processed inside a satellite receiver. Multiple block downconversion receivers are capable of independently selecting channels because each can process the entire block of signals.

BNC Connector
A weatherproof twist lock coax connector standard on commercial video equipment and used on some brands of satellite receivers.

Boresight
That direction along the principle axis of either a transmitting or a receiving antenna.

Buttonhook Feed
A rod shaped like a question mark supporting the feedhorn and LNA. A buttonhook feed for use with commercial grade antennas is often a hollow waveguide that directs signals from a feedhorn to an LNA behind the antenna.

CATV
An abbreviation for Community Antenna Television - another name for cable TV.

CCD
Charge coupled device. Used in MAC transmissions for temporarily storing video signals.

C-Band
The 3.7 to 4.2 GHz band of frequencies at which some broadcast satellites operate.

Carrier
A radio signal of a single frequency that is modulated to carry information.

Carrier-to-Noise Ratio (C/N)
The ratio of the received carrier power to the noise power in a given bandwidth. The C/N is an indicator of how well an earth receiving station will perform in a particular location, and is calculated from satellite power levels, antenna gain and the combined antenna and LNA noise temperature.

Carrier-to-Noise Ratio
The C/N expressed in decibels per Hertz of signal bandwidth.

Cassegrain Feed System

An antenna feed design that includes a primary reflector, the dish, and a secondary reflector which redirects microwaves via a waveguide to a low noise amplifier.

Channel

A segment of bandwidth used for one complete communication link.

Chrominance

The hue and saturation of a color. The chrominance signal is modulated onto a 3.58 MHz carrier.

Chrominance Signal

The color component of the composite baseband video signal assembled from the I and Q portions. Phase angle of the signal represents hue and amplitude represents color saturation.

Circular Polarization

Electromagnetic waves whose electric field uniformly rotates along the signal path. Broadcasts used by Intelsat and other international satellites use circular not horizontally or vertically polarized waves as are common in North American and European transmissions.

Clark Belt

The circular orbital belt at 22,247 miles above the equator, named after the writer Arthur C. Clarke, in which satellites travel at the same speed as the earth's rotation. Also called the geosynchronous or geostationary orbit.

Color Bars

A test pattern of specifically colored vertical bars used as a reference to test the performance of a color television.

Color Sync Burst

A "burst" of 8 to 11 cycles of the 3.579545 MHz color subcarrier frequency. This waveform is located on the back porch of each horizontal blanking pulse during color transmissions. It serves to synchronize the color subcarrier's oscillator with that of the transmitter in order recreate the raw color signals.

Coaxial Cable

A cable for transmitting high frequency electrical signals with low loss. It is composed of an internal conducting wire surrounded by an insulating dielectric which is further protected by a metal shield.

Composite Baseband Signal

The complete audio and video signal without a carrier wave. Satellite signals have audio baseband information ranging in frequency from zero to 3400 Hertz. NTSC video baseband is from zero to 4.2 MHz.

Composite Video Signal

The complete video signal consisting of the chrominance and luminance information as well as all sync and blanking pulses.

Compounding

A form of noise reduction using compression at the transmitting end and expansion at the receiver. A compressor is an amplifier that increases its gain for lower power signals. The effect is to boost these components into a form having a smaller dynamic range. A compressed signal has a higher average level, and therefore, less apparent loudness than an uncompressed signal, even though the peaks are no higher in level. An expander reverses the effect of the compressor to restore the original signal.

Contrast

The ratio between the dark and light areas of a television picture.

Cross Polarization

Term to describe signals of the opposite polarization to another being transmitted and received. Cross-polarization discrimination refers to the ability of a feed to detect one polarity and reject the opposite polarity signals.

Crosstalk

Interference between adjacent channels.

DC Power Block

A device which stops the flow of DC power but permits passage of higher frequency signals.

Decibel (dB)

A term that expresses the ratio of power levels used to indicate gains or losses of signals. Decibels relative to one watt, milliwatt and millivolt are abbreviated as dBw, dBm and dBmv, respectively.

Declination Offset Angle

The adjustment angle of a polar mount between the polar axis and the plane of a satellite antenna used to aim at the geosynchronous arc.

Decoder

A circuit that restores a signal to its original form after it has been scrambled.

De-emphasis

A reduction of the higher frequency portions of an FM signal used to neutralize the effects of pre-emphasis. When combined with the correct level of pre-emphasis, it reduces overall noise levels and therefore increases the signal-to-noise ratio.

Demodulator

A device which extracts the signal from the transmitted carrier wave.

Detent Tuning

Tuning into a satellite channel by selecting a preset resistance.

Digital

Describes a system or device in which information is transferred by electrical "on-off," "high-low," or "1/0" pulses instead of continuously varying signals as in an analog message.

Digital-to-Analog Converter

A circuit that converts digital signals into their equivalent analog form.

Direct Broadcast Satellite (DBS)
A term commonly used to describe satellite broadcasts directly to homes using the Ku-band, roughly in the downlink frequency range from 11 to 13 GHz.

Dish
Jargon for a parabolic microwave antenna.

Downconverter
A circuit that lowers the high frequency signal to a lower, intermediate range. There are three distinct types of downconversion used in satellite receivers: single downconversion; dual downconversion; and block downconversion.

Downlink Antenna
The antenna on-board a satellite which relays signals back to earth.

Drifting
An instability in a preset voltage, frequency or other electronic circuit parameter.

Dual Feedhorn
A feedhorn which can simultaneously receive both horizontally and vertically polarized signals.

Earth Station
A complete satellite receiving or transmitting station including the antenna, electronics and all associated equipment necessary to receive or transmit satellite signals.

Effective Isotropic Radiated Power (EIRP)
A measure of the signal strength that a satellite transmits towards the earth below. The EIRP is highest at the center of the beam and decreases at angles away from the boresight.

Elevation Angle
The vertical angle measured from the horizon up to a targeted satellite.

Equalizing Pulses
A series of pulses occurring before and after the serrated vertical sync pulse to ensure proper interlacing. The equalizing pulses are inserted at twice the horizontal scanning frequency.

FCC
The Federal Communications Commission, the regulatory board which sets standards for communications within the United States.

f/D Ratio
The ratio of an antenna's focal length to diameter. It describes antenna "depth."

Feedhorn
A device that collects microwave signals reflected from the surface of an antenna. It is mounted at the focus of all prime focus parabolic antennas.

Field
One half of a complete TV picture or frame, composed of 262.5 scanning lines. There are 60 fields per second for black/white TV and 59.94 fields per second for color TV in NTSC transmission.

Focal Length
The distance from the reflective surface of a parabola to the point at which incoming satellite signals are focused, the focal point.

Footprint
The geographic area towards which a satellite downlink antenna directs its signal. The measure of strength of this footprint is the EIRP.

Forward Error Correction
FEC is a technique for improving the accuracy of data transmission. Excess bits are included in the out-going data stream so that error correction algorithms can be applied upon reception.

Frame
One complete TV picture. It is composed of two fields and has a total of 525 scanning lines in NTSC transmission.

Frequency
The number of vibrations per second of an electrical or electromagnetic signal expressed in cycles per second or Hertz.

Front Porch
The portion of the horizontal blanking pulse that precedes the horizontal sync pulse.

Gain
The amount of amplification of input to output power often expressed as a multiplicative factor or in decibels.

Gain-to-Noise Temperature Ratio (G/T)
The figure of merit of an antenna and LNA. The higher the G/T the better the reception capabilities of an earth station.

Geostationary Orbit
See Clarke Belt.

GigaHertz (GHz)
Billions of cycles per second.

Global Beam
A footprint pattern used by communication satellites targeting nearly 40% of the earth's surface below. Many Intelsat satellites use global beams.

Ground Noise
Unwanted microwave signals generated from the warm ground and detected by a dish.

Hall Effect Sensor
A device used to count the number of rotations of an actuator motor. A small current carrying wire generates voltage pulses when passed by a permanent magnet.

Hardline
A low-loss coaxial cable that has a continuous hard metal shield instead of a conductive braid around the outer perimeter.

Heliax

A thick low-loss cable used at high frequencies also known as hardline.

High Definition Television (HDTV)

An innovative television format having approximately twice the number of scan lines in order to improve picture resolution and viewing quality.

Horizontal Blanking Pulse

The pulse that occurs between each horizontal scan line and extinguishes the beam illumination during the retrace period.

Horizontal Sync Pulse

A 5.08 microsecond rectangular pulse riding on top each horizontal blanking pulse. It synchronizes the horizontal scanning at the television set with that of the television camera.

I Signal

One of the two color video signals which modulate the color subcarrier. It represents those colors ranging from reddish orange to cyan.

Impulse Pay-per-view

Impulse pay-per-view (IPPV) is a feature of a decoder that allows an authorized subscriber to purchase a one-time scrambled program at will. IPPV shows are selected by a button on the decoder or its remote control unit.

Inclinometer

An instrument used to measure the angle of elevation to a satellite from the surface of the earth.

Interlaced Scanning

A scanning technique to minimize picture flicker while conserving channel bandwidth. Even and odd numbered lines are scanned in separate fields both of which when combined paint one frame or complete picture.

Intermediate Frequency (IF)
A middle range frequency generated after downconversion in any electronic circuitry including a satellite receiver.

INTELSAT
The International Telecommunication Satellite Consortium, a body of 154 countries working towards a common goal of improved worldwide satellite communications.

Isolator
A device that allows signals to pass unobstructed in one direction but which attenuates their strength in the reverse direction.

Kelvin Degrees (K)
The temperature above absolute zero, the temperature at which all molecular motion stops, graduated in units the same size as degrees Celcius (°C). Absolute zero equals -273 °C or -459 °F.

Ku-Band
The microwave frequency band between approximately 11 and 13 GHz.

LC Modulator
A modulator constructed from passive inductive and capacitive circuit elements.

Latitude
The measurement of a position on the surface of the earth north or south of the equator measured in degrees of angle.

Line Amplifier
An amplifier in a transmission line that boosts the strength of a signal.

Local Oscillator
A device used to supply a stable single frequency to an upconverter or downconverter. The local oscillator signal is mixed with the carrier wave to change its frequency.

Longitude
The distance east or west of the prime meridian as measured in degrees.

Low Noise Amplifier (LNA)
A device that receives and amplifies the weak satellite signal reflected by an antenna via a feedhorn. LNAs have noise temperature rated in degrees Kelvin.

Low Noise Block Downconverter (LNB)
An LNA which also downconverts the whole 500 or 750 MHz satellite bandwidth at once to an intermediate frequency range.

Low Noise Converter (LNC)
An LNA and a conventional downconverter housed in one weatherproof box.

Magnetic Variation
The difference between true north and the north indication of a compass.

Master Antenna TV (MATV)
Broadcast receiving stations that use one or more high-quality centrally located UHF and/or VHF antennas which relay their signals to many televisions in a local apartment/condo or group-housing complex.

MegaHertz (MHz)
Millions of cycles per second.

Microprocessor
The central processing unit of a computer, either on a single integrated (IC) circuit chip or on several ICs.

Microwave
The frequency range from approximately 500 MHz to 30 GHz and above.

Modulation
A process in which a message is added to a carrier wave. Among other methods, this can be accomplished by frequency or amplitude modulation, known as AM or FM, respectively.

Monochrome
A black and white television picture.

Mount
The structure that supports an earth station antenna. Polar and az-el mounts are the most common variety.

Multiple Analog Component (MAC) Transmissions
An innovative television transmission method which separates the data, chrominance and luminance components and compresses them for sequential relay over one television scan line. There are presently five systems being designed: A-MAC; B-MAC; C-MAC; D-MAC; and E-MAC.

Multiplexing
The simultaneous transmission of two or more signals over a single communication channel. The interleaving of the luminance and chrominance signals is one form of multiplexing, known as frequency multiplexing. MAC transmissions make use of time division multiplexing.

N-Connector
A low-loss coaxial cable connector used at the elevated C-band microwave frequencies.

NTSC
The National Television Standards Committee which created the standard for North American TV broadcasts.

NTSC Color Bar Pattern
The standard test pattern of six adjacent color bars including the three primary colors plus their three complementary shades.

Negative Picture Phase

Positioning the composite video signal so that the maximum level of the sync pulses are at 100% amplitude. The brightest picture signals are in the opposite negative direction.

Negative Picture Transmission

Transmission system used in North America and other countries in which a decrease in illumination of the original scene causes an increase in percentage of modulation of the picture carrier. When demodulated, signals with a higher modulation percentage have more positive voltages.

Noise

An unwanted signal which interferes with reception of the desired information. Noise is often expressed in degrees Kelvin or in decibels.

Noise Figure

The ratio of the actual noise power generated at the input of an amplifier to that which would be generated in an ideal resistor. The lower the noise figure, the better the performance.

Noise Temperature

A measure of the amount of thermal noise present in a system or a device. The lower the noise temperature, the better the performance.

Odd Field

The half frame of a television scan which is composed of the odd numbered lines.

Offset Feed

A feed which is offset from the center of a reflector for use in satellite receiving systems. This configuration does not block the antenna aperture.

PAL

Phase Alternate Line. The European color TV format which evolved from the American NTSC standard.

Pad

A concrete base upon which a supporting pole and antenna can be mounted.

Parabola

The geometric shape that has the property of reflecting all signals parallel to its axis to one point, the focal point.

Pay-per-view

Pay-per-view is a method of purchasing programming on a per-program basis.

Persistence of Vision

The physiological phenomena whereby a human eye retains perception of an image for a short time after the image is no longer visible.

Phase

A measure of the relative position of a signal relative to a reference expressed in degrees.

Phase Distortion

A distortion of the phase component of a signal. This occurs when the phase shift of an amplifier is not proportional to frequency over the design bandwidth.

Picture Detail

The number of picture elements resolved on a television picture screen. More "crisp" pictures result as the number of picture elements is increased.

Polar Mount

An antenna mount that permits all satellites in the geosynchronous arc to be scanned with movement of only one axis.

Polarization

A characteristic of the electromagnetic wave. Four senses of polarization are used in satellite transmissions: horizontal; vertical; right-hand circular; and left-hand circular.

Positive Picture Phase

Positioning of the composite video signal so that the maximum point of the sync pulses are at zero voltage. The brightest illumination is caused by the most positive voltages.

Pre-Emphasis

Increases in the higher frequency components of an FM signal before transmission. Used in conjunction with the proper amount of de-emphasis at the receiver, it results in combatting the higher noise detected in FM transmissions.

Primary Colors

Red, green and blue.

Prime Focus Antenna

A parabolic dish having the feed/LNA assembly at the focal point directly in the front of the antenna.

Q Signal

One of two color video signal components used to modulate the color subcarrier. It represents the color range from yellowish to green to magenta.

Radio Frequency

The approximately 10 KHz to 100 GHz electromagnetic band of frequencies used for man-made communication.

Raster

The random pattern of illumination seen on a television screen when no video signal is present.

Reed Switch

A mechanical switch which uses a reed to make and break electrical contact and thus to count pulses which are sent to the antenna actuator controller.

Reference Signal

A highly stable signal used as a standard against which other variable signals may be compared and adjusted.

Retrace

The blanked-out line traced by the scanning beam of a picture tube as it travels from the end of any horizontal line to the beginning of either the next horizontal line or field.

SAW (Surface Acoustic Wave) Filter

An electronic device which allows a sharp transition between regions of allowed and attenuated frequencies.

Satellite Receiver

The indoors electronic component of an earth station which downconverts, processes and prepares satellite signals for viewing or listening.

Scanning

The organized process of moving the electron beam in a television picture tube so an entire scene is drawn as a sequential series of horizontal lines connected by horizontal and vertical retraces.

Scrambling

A method of altering the identity of a video or audio signal in order to prevent its reception by persons not having authorized decoders.

Screening

Use of metal to screen out unwanted TI.

Serrated Vertical Pulse

The television vertical sync pulse which is subdivided into six serrations. These sub-pulses occur at twice the horizontal scanning frequency.

Servo Hunting

An oscillatory searching of the feedhorn probe when use of inadequate gauge control cables results in insufficient voltage at the feedhorn.

Side Lobe

A construct used to describe an antenna's ability to detect off-axis signals. The larger the side lobes, the more noise and interference an antenna can detect.

Single Channel Per Carrier (SCPC)
A satellite transmission system that employs a separate carrier for each channel, as opposed to frequency division multiplexing that combines many channels on a single carrier.

Signal-to-Noise Ratio (S/N)
The ratio of signal power to noise power in a specified bandwidth, usually expressed in decibels.

Skew
A term used to describe the adjustment necessary to fine tune the feedhorn polarity detector when scanning between satellites.

Sparklies
Small black and/or white blips or dots in a television picture indicating an insufficient signal-to-noise ratio. Also known as "snow."

Spherical Antenna
An antenna system using a section of a spherical reflector to focus one or more satellite signals to one or a series of focal areas.

Splitter
A device that takes a signal and splits it into two or more identical but lower power signals.

Subcarrier
A carrier wave that modulates another higher frequency carrier. In satellite transmissions a 6.8 MHz audio subcarrier is often used to modulate the C-band carrier. In television, a 3.58 MHz subcarrier modulates the video carrier on each channel.

Surface Acoustic Wave
A sound or acoustic wave travelling on the surface of the optically polished surface of a piezoelectric material. This wave travels at the speed of sound but can pass frequencies as high as several gigahertz. See SAW Filter.

Synchronizing Pulses
Pulses imposed on the composite baseband video signal used to keep the television picture scanning in perfect step with the scanning at the television camera.

TVRO
A television receive-only earth station designed only to receive but not to transmit satellite communications.

Terrestrial Interference (TI)
Interference of earth-based microwave communications with reception of satellite broadcasts.

Thermal Noise
Random, undesired electrical signals caused by molecular motion, known more familiarly as noise.

Trace
The movement of the electron beam from left to right on a television screen.

Threshold
A minimal signal to noise input required to allow a video receiver to deliver an acceptable picture.

Transponder
A microwave repeater, which receives, amplifies, downconverts and retransmits signals at a communication satellite.

Trap
An electronic device that attenuates a selected band of frequencies in a signal.

UHF
Ultrahigh frequencies ranging from 300 to 3,000 MHz.

Upconverter
A device that increases the frequency of a transmitted signal.

Uplink

The earth station electronics and antenna which transmits information to a communication satellite.

VSWR (Voltage Standing Wave Ratio)

A measure of the percentage of reflected power to the total power impinging upon a device.

Vertical Blanking Pulse

A pulse used during the vertical retrace period at the end of each scanning field to extinguish illumination from the electron beam.

Vertical Sync Pulse

A series of pulses which occur during the vertical blanking interval to synchronize the scanning process at the television with that created at the studio. See also Serrated Vertical Pulse.

VHF

Very high frequency range from 30 to 300 MHz

Video Signal

That portion of the transmitted television signal containing the picture information.

Voltage Tuned Oscillator (VTO)

An electronic circuit whose output oscillator frequency is adjusted by voltage. Used in downconverters and satellite receivers to select from among transponders.

Video Monitor

A television that accepts unmodulated baseband signals to reproduce a broadcast.

APPENDIX E. REFERENCE MATERIALS

Satellite TV Guides

Channel Guide
300 East Hampden, Suite 340
Englewood, CO 80110
(303) 654-3006

On Sat
Triple D Publishing
P.O. Box 2384
Shelby, NC 28151
(800) 438-2020

Orbit
CommTek Publishing
P.O. Box 1700
Hailey, ID 83333
(800) 792-5541

Satellite Dish Magazine
P.O Box 8
Memphis, TN 38101
(901) 521-1580

Satellite TV Weekly
Fortuna Communications
P.O. Box 308
928 Main Street
Fortuna, CA 95540
(800) 556-8787

Satellite Entertainment Guide
Vogel and Son Publishing Company
P.O. Box 8266
Edmonton, Alberta T6H 4P1
Canada
(403) 425-1169

Trade Publications

Coops Satellite Digest
Triple D Publishing
P.O Box 2384
Shelby, NC 28151-2384
(704) 482-9673

Satellite Marketing
CES Publishing
345 Park Avenue South
New York, NY 10010
(212) 686-7744

Private Cable
Weisner Publishing Company
5951 South Middlefield Road
Littleton, CO 80123
303) 798-1274

Satellite Communications
Cardiff Publishing
6430 S. Yosemite Street
Englewood, CO 80111
(303) 694-1522

Satellite Dealer
CommTek Publishing
P.O. Box 53
Boise, ID 83707
(208) 322-2800

Satellite Retailer
Triple D Publishing
P.O. Box 2384
Shelby NC 28151
(704) 482-9673

Satellite TV Opportunities
1717 East University Avenue
Oxford, MS 38655
(601) 236-5510

Satellite Times
Triple D Publishing
501 North Washington Street
Shelby, NC 28150
(704) 482-9673

STV Magazine
Triple D Publishing
501 N. Washington Street
Shelby, NC 28150
(704) 482-9673

Signal Magazine
Fernwood Publishing
P.O. Box 238, Station D
Scarborough, Ontario M1R 5B7
Canada
(416)759-6639

TVRO Technology
Weisner Publications
5951 South Middlefield Road
Littleton, CO 80123
(303) 798-1274

APPENDIX F. MANUFACTURERS

American Extasy
P.O. Box 1948
New York, NY 10156
(212)696-4111

Conifer
1400 North Roosevelt Avenue
Burlington, IA 52601
(319)752-3607

Fun Channel
Space Age Video of Texas
One Summit Avenue, Suite B-103
Fort Worth, TX 76102
(817)332-7294

General Instruments Corporation
2200 ByBerry Road
Hatboro, PA 19040
(215)674-4800

Hamlin
13610 First Avenue South
Seattle, WA 98168
(206)246-9330

Microwave Filter Company
6743 Kinne Street
East Syracuse, NY 13057
(800)525-5571

M/A COM
Cable Com Group
P.O. Box 1729
Hickory, NC 28603
(800)438-3331

Link-A-Bit
(619)457-2340

Oak Communications
Satellite Systems Division
P.O. Box 517
Crystal Lake, IL 60014
(815)459-5000

Oak Communications
Cable Division
16935 West Bernardo Drive
Rancho Bernardo, CA 92127
(619)485-9880

Pico Products
103 Commerce Blvd.
Liverpoll, NY 13088
(800)822-7420

Scientific Atlanta
4356 Communications Drive
Norcross, GA 30093
(404)925-5778

Scientific Atlanta—mailing address
P.O. Box 105027
Atlanta, GA 30348

Zenith Electronics Corporation
1000 Milwaukee Avenue
Glenview, IL 60025
(312)391-8338

Tekscan
1440 Goodyear Drive
El Paso, TX 79936
(915) 594-3555

Our Best To You

A MATE FOR YOUR HOME SATELLITE TV INSTALLATION & TROUBLESHOOTING MANUAL

Newly Released

(Available in English & Spanish)

$39.95

The Home Satellite TV Installation Videotape

"Every dealer in America should stock this tape in their showroom as a sales tool."
Scott Zimmer
International Division
Echosphere

"An invaluable tool for apprentice installers and do-it-yourselfers. Its low cost will more than pay for itself in time saved."
Sergio Murillo
Televisat, Mexico City

"This book . . . puts it all in one place and will make it possible to throw out that old collection."
Taylor Howard
Director of Research
Chaparral Communications

"This book will revolutionize the industry by allowing the consumer to really understand satellite TV."
Keith Lamonica
FM America

Understand Satellite TV

Installing and "Tuning Up" Your Satellite TV System

- Performing a Site Survey
- Detecting & Avoiding TI
- Installing All Equipment
- Tracking the Arc

AVAILABLE FROM
BAYLIN/GALE PRODUCTIONS
1905 MARIPOSA — SUITE 101
BOULDER, CO 80302
(303) 449-4551
Please add $2 to the cost of every item shipped—See Order Form below

REFERENCE MATERIAL ORDER FORM

Book	Number of Copies	Unit Cost	Total Cost
Ku-Band Satellite TV Theory, Installation and Repair		$19.95	
The Home Satellite TV Installation Video Tape (English/Spanish/VHS/Beta)		$39.95	
The Home Satellite TV Installation and Troubleshooting Manual (English/Spanish)		$29.95	
The Home Satellite TV Installation and Troubleshooting Manual and Installation Video Tape Combo		$59.95	
Satellites Today - The Complete Guide to Satellite Television		$ 9.95	
ASTI Manual - The Avoidance/Suppression Approach to Eliminating Terrestrial Interference at TVRO Earth Stations		$59.00	
The Hidden Signals on Satellite TV		$14.95	

$2.00 per book shipping cost ________

Total Cost $______

Name ____________________

Company ____________________

Address ____________________

____

____

Telephone ____________________

Please send money order or check to:

Baylin/Gale Productions
Suite 101
1905 Mariposa Street
Boulder, Colorado 80302
(303) 449-4551

Orders from Canada, Mexico and overseas countries must be money orders in U.S. dollars or cash. Please enclose the required amount for shipping either by air or sea.

APPENDIX G

Scrambling Technique	Scrambling Depth	Video Security	Residual Effects in Descrambled Video	Descrambler Hardware Complexity	Cost
RF METHODS					
1. Tone Jammer	Marginal. Scrambles Audio also	Inadequate	Useful luminance energy lost	Low. 1 Trap per scrambled channel	Low
2. Video Inversion	Marginal	Inadequate	Luminance and chrominance distortions due to imperfect carrier recovery	Complex	High
3. Sinewave Sync Suppression	Adequate	Adequate	Noise transfer from descrambling signal to video	Low	Low
4. Squarewave Sync Suppression	Adequate	Inadequate (sync easily restored)	Video jitter due to inaccurate timing	Low	Low
5. Frequency Inversion	Good. Scrambles Audio also	Good	Scrambled picture due to inaccurate timing	Moderate	Moderate
6. Nonlinear Filter	Good	Good	Distortions due to filter mismatch	Low	Low

Table G-1. Selected Video Scrambling RF Methods.

Scrambling Technique	Scrambling Depth	Video Security	Residual Effects in Descrambled Video	Descrambler Hardware Complexity	Cost
BASEBAND METHODS					
1. Video Inversion / Sync Suppression	Adequate	Adequate	Distortions due to inaccurate DC restoration	Moderate	Moderate
2. Video Jitter	Good	Excellent	Jittered video due to inaccurate timing	Moderate	High
3. Line Reversals	Good	Adequate	Negligible	Low	Moderate
4. Line Permutations	Excellent	Excellent	Negligible	High	High
5. Line Dicing	Excellent	Excellent	Significant segment distortions in CATV links due to VSB filtering and multipath	High	High
6. MAC A, B or C	Good - in conjunction with other scrambling methods	Good	Not presently applicable to 6MHz CATV links	Not Known	Not Known

Table G-2. Video Scrambling Baseband Methods.

TABLE G-3. C & KU-BAND SATELLITE RECEIVERS SUCCESSFULLY TESTED WITH VIDEOCIPHER II. *(Courtesy of General Instruments).*

Locate the name and model of your satellite receiver below. The column heading will tell you which installation procedure (Composite, 70 MHz IF or 70 MHz IF Loop-through) to follow. If you don't find the name of your receiver, contact the satellite receiver manufacturer for additional information. Units marked with an asterisk(*) require a factory authorized modification or adjustment for proper operation with the VideoCipher II descrambler. Contact the satellite receiver manufacturer for additional information.

70 MHz IF Connection

Amplica CSR 100
Amplica CSR 200
Amplica CSR 300
Anderson Scientific 910 HB
Anderson Scientific 1010 HB
Anderson Scientific 2010 HB
Avcom COM-2A (LNC RDC-11A)
Avcom COM-2B (LNC RDC-11A)
Avcom COM-3 (LNC RDC-3)
Avcom COM-3R (LNC RDC-3)
Birdview MR 20/20*
Birdview 20/20 M2*
Birdview 20/20 M3*
Birdview 20/20
Boman SR-1200
Boman SR-1500
Boman SR-2500
Channel Master 6129
Channel Master 6130
Channel Master 6131
Channel Master 6134
Channel Master 6138
Chaparral Sierra II
Drake ESR 24
Drake ESR 224
Drake ESR 240
Drake ESR 240A
Drake ESR 324
Drake Black Widow Series 1
Earth Terminals
GCI 8300
Intersat Baby Q
KLM Skyeye X
Lowrance System 70X
Luxor 9550
ProSat 315 (LNC P-511)
ProSat 330
SatStar Elan
Sat Tec R-5000
Sigma-Vu Mark 11A
STS MBS-SR
Toki TR220
Uniden UST-1000
Uniden UST-3000
USS-MASPRO SR-2
Vector BSR-4000
Wilson YM400
Wilson YM450
Wilson YM1000
Winegard RF-1000

Composite Video Connection

Amway ASR 2000
Arcfinder 2000
Brooks B-250
Brooks B-250A
BSR SX2010
Channel Master 6135
Channel Master 6137
Channel Master 6144
Chaparral Cheyenne
Chaparral Sierra II
Cincinnati Microwave STR500
Conifer XT-100
Conifer XT-200
Drake ESR 424 Block System
Drake ESR 524 Block System
Drake ESR-9241
DX Communications DSB-600
DX Communications DSB-600A
DX Communications DSB-700
DX Communications DSB-700A
DX Communications DSB-700S*
DX Communications DSB-800
General Instrument SRX-207
Hytek/Dexcel SRX-500 plus
Janeil BCR-2000*
Janeil BCR-5000
Kenwood KSR-1000*
Luxor 9570 - Mark II
Luxor 9900
Luxor 9995
M/A-COM H-1
M/A-COM T-1
M/A-COM T-2
M/A-COM T-6
NORSAT JR-100
NORSAT JR-200
NORSAT JR-300*
Pansonic C-1000
Panasonic C-2000
Panasonic C-2000A
Panasonic Ku/C-6000
Radio Shack SR2010
Radio Shack TDP 900
Sat Tec R-5100*
Scientific Atlanta Homesat 800
Standard MT800-Agile Omni
Uniden UST-5000*
Uniden UST-6000*
Uniden UST-7000*
USS-MASPRO SR-3
Zenith ZS-3000
Zenith ZS-4000

70 MHz IF Loop-Through Connection

Amway ASR-2000
Anderson Scientific ST 4010
Chaparral Sierra
Drake ESR 324 Block System
Drake ESR 324S Block System
Drake ESR 424 Block System
Drake ESR 524 Block System
Drake Black Widow Series II
Houston Tracker System V
Pico CR 1000
Pico HR 1000
Sigma -Vu Mark III
Sigma-Vu Mark V
Uniden UST - 5000
Uniden UST - 6000
Uniden UST - 7000

Figure G-1. Diagnostic Data Screen Using Setup 1.

Figure G-2. Diagnostic Data Screen Using Setup 0.

VIDEOCIPHER II DIAGNOSTICS

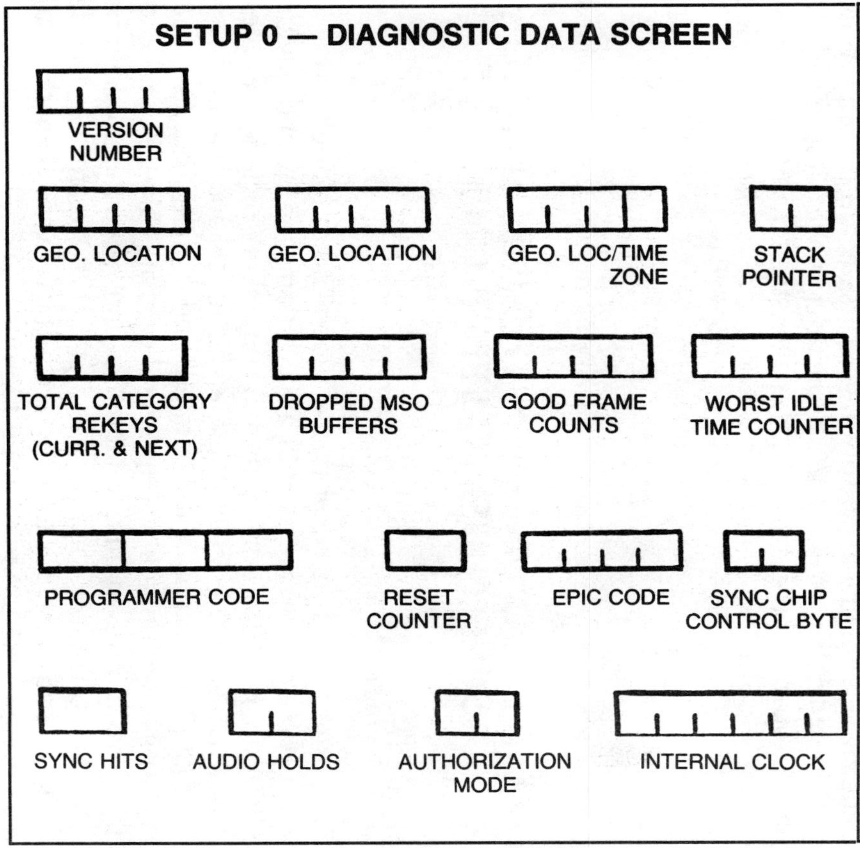

SETUP 0 — DIAGNOSTIC DATA SCREEN

VERSION NUMBER

GEO. LOCATION

GEO. LOCATION

GEO. LOC/TIME ZONE

STACK POINTER

TOTAL CATEGORY REKEYS (CURR. & NEXT)

DROPPED MSO BUFFERS

GOOD FRAME COUNTS

WORST IDLE TIME COUNTER

PROGRAMMER CODE

RESET COUNTER

EPIC CODE

SYNC CHIP CONTROL BYTE

SYNC HITS

AUDIO HOLDS

AUTHORIZATION MODE

INTERNAL CLOCK

Table G-5. Setup 0.

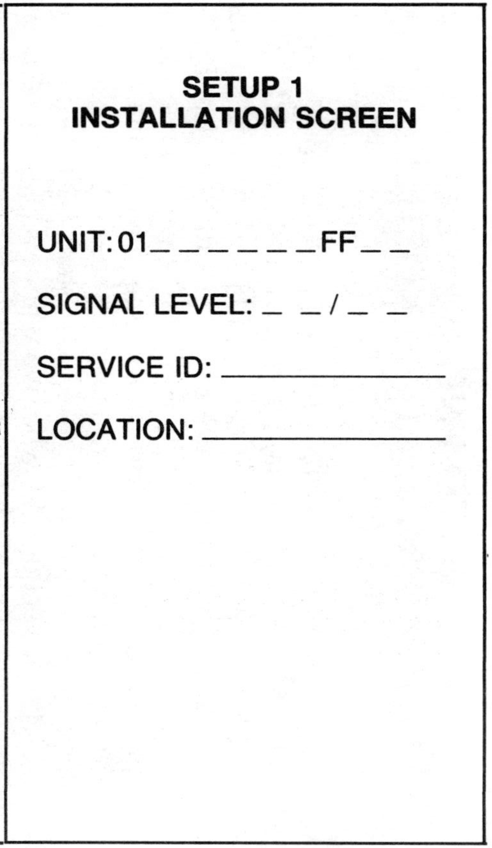

SETUP 1
INSTALLATION SCREEN

UNIT: 01_ _ _ _ _ _ _FF_ _

SIGNAL LEVEL: _ _/_ _

SERVICE ID: ______________

LOCATION: ______________

Table G-4. Setup 1.

GETTING TO KNOW YOUR VIDEOCIPHER II DIAGNOSTICS

As more and more satellite systems are installed through-out the United States, Mexico and Canada, we are faced with the fact that the VideoCipher II encoding system is going to be the defacto standard just as the NTSC television standard is used in North America. Be that as it may if we want to watch quality programing we will need to buy a VideoCipher II decoder for our Home or commercial Satellite System. Many people need to be kept informed as to how a particular piece of electronic equipment operates, like your stereo, VCR, and even your VideoCipher II decoder. So what we will cover in this section is the diagnostic screens that can be displayed by selecting the appropriate menu. This will tell us if your VC II decoder is working like it was designed to operate. The two major diagnostic screens that can be brought up are SET-UP 1 and SET-UP 0. There is also SET-UP 2,3, AND 4. But 1 and 0 are what we will consentrate on.

WHAT TO LOOK FOR ON THE INSTALLATION SCREEN USING SETUP ONE

1. ROW 1—The UNIT ID NUMBER.

This is the address number of the VideoCipher module. This 12 digit alpha-numeric number should always start with 01 and have FF at the 9th and 10th positions. Any other numbers or letters in these positions would identify the module to be faulty and should be replaced. It can be thought of like a telephone number, because each module has its own unique 12 digit number. This hexa-decimal code gives the module its own unique identifier and when the programmer sends this authorization code in the digital data stream the decoder will see it and cause the box to authorize just like when someone dials your telephone number and your phone rings.

2. ROW 2—SIGNAL LEVEL

This level can be 50/50. General Instruments states that below 45/45 is not acceptable. However, video has been achieved with numbers as low as 1/5. The pictures will be of poor quality and the sound will be noisy but the videocipher module will turn on in some instances. There are some satellite receiver manufactures that claim to be VC II compatible but will not turn on even when a signal level of 50/50 is received. Thus signal level is only one indication of whether a VC II / receiver combination will work.

The first number is the instantaneous level and the second number is the average. These two numbers should be similar once the channel has been tuned in for several minutes. If they are drastically different, use the instantaneous value. Adjusting the receivers baseband or composite video output level will cause the numbers to go up or down as will adjusting the VC II composite video level input. (If using the receivers 70 mhz loop through, this composite control will have no effect.

If a low signal level (less than 30/30) is accompanied with poor picture quality, and site problems (terrestial interference, or poor dish alignment) have been excluded, check the signal levels on other videocipher II channels, whether authorized or not. If they too are bad, you may have a faulty module or you may want to recheck for site problems. If they are good, the problem may be with the individual channel or programmer, and not the module.

3. ROW 3—The SERVICE ID

This simply identifies whether or not the module has been authorized. NONE or AO indicates not authorized. AO, accompanied by a black screen when the "VIEW" button is selected, means the programmer is not offering subscriptions to that channel. Any other code represents that authorization has been received by the module.

4. ROW 4—The LOCATION

Either SET or NOT SET, provides information similar to the SERVICE ID concerning the authorization status of the module.

WHAT TO LOOK FOR ON THE DIAGNOSTIC DATA SCREEN: (SETUP 0)

1. GOOD FRAME COUNT

The only numbers of any real importance when troubleshooting are found in the third column second row. This is the GOOD FRAME COUNT number. The speed at which the numbers change should be equal to the internal clock of the microprocessor found in row 4 of column 4. This should count at about 500 counts per 90 seconds, if it is not there is a problem with the strength or quality of the signal being received by the module. If the satellite receiver is truly VC II compatible than this number will be steadily incrementing. If it is counting in bursts or is not moving at all than chances are that the composite video level input to the decoder is wrong and needs to be adjusted or the satellite receiver is not VC II compatible. The frame count numbers will also stop when terrestial interference is present due to increased noise.

Other site problems can be dish mis-alignment, a faulty receiver or a poor connection, bad patch cord between the receiver and the videocipher II decoder or between the LNB and receiver. The good frame counts counter can be reset to zero by pressing the "ENTER" key.

2. The AUTHORIZATION MODE

This can be one of several of the following codes and can be found in the third column fourth row, this will indicate the status of the VC II module:

The letters will be F for (fixed module scrambling), which is where every VC II module will turn on or S for (addressable scrambling), where only authorized units will turn on.

The right letter will be A for (unit authorized), B for (unit blacked out), This is determined by your phone number area code or M for (authorization missing) as is the case with a new decoder that has not been authorized.

SM = The unit has never been authorized. This could stem from a site problem or lack of authorization being a new unit or no authorization signal sent or received from the programmer.

SL = The current program rating exceeds that limit set in the unit. This can be changed by pressing setup 3 and raising the unit to a higher level if a password is not programmed.

SB = The unit has been authorized, however not for this particular program.

SA = The unit has been authorized for this program and should be received.

FA = The programmer is transmitting a blanket authorization to all VC II modules, regardless of subscription status. This is often done at intermittent times during test scrambling and may include a "barker" screen across the picture providing subscription advertisements.

3. The AUDIO HOLDS

This is found on row 2 line 4, and should be one of two types of codes. When a program is authorized, a 00 or 01 should appear. When a program is not authorized, a 2C, 2D, 2E or 2F should appear. Any other codes would identify a module audio problem and the module would need to be replaced.

4. The EPIC CODE

Which is found on row 3 line 3 represents a master code which is updated every month by the programmer. It's important that the epic code on the diagnostic screen matches the current months epic code. This problem can occur if the receiver is not plugged in during the epic change or if the decoder is unable to receive the new code. In some instances, the programmer's inability to hit all the authorized modules, due to weather conditions or equipment failures, may inhibit the modules ability to receive the epic change. If you feel that the epic code may not reflect the current code, contact the programmer to determine it's current status. This problem will generally occur the first few days of a given month since this is when epic changes are made. A module that has never been authorized will have "CBDO" in this location.

5. The TOTAL CATEGORY REKEYS

Is located in row 1 line 2 this will reflect the number of "hits" the Videocipher has received since it's authorization. This number can be reset to zero by pressing the "ENTER" key. Therefore the actual number is not important. The prime concern with this number is that it should increment. This count should be increasing every 3 to 6 hours depending on the number of subscribers and when the programmer hits your module. If these numbers do not reflect a change over this period of time, either the programmer is not updating the module or the module is unable to receive the authorization "hits". This could be caused by site problem or interference. The programmer should be notified so the module address number can be put on a "hit" list. Obviously, the receiver needs to be tuned to the particular satellite channel in question. If the hit is not received within 6 hours, you can assume that the module is faulty if good video is received on other unscrambled channels.

6. The PROGRAMMER CODE

This code is located in row 1 line 3. This indicates the programmers ID number, example CNN, HBO, ESPN etc.

7. The top row of numbers indicate the geographical location of the decoder after it has been authorized. Before authorization these numbers will be all zero's.

I hope this insight into the inner workings of the diagnostic screens will be of help in diagnosing any site or decoder problems.

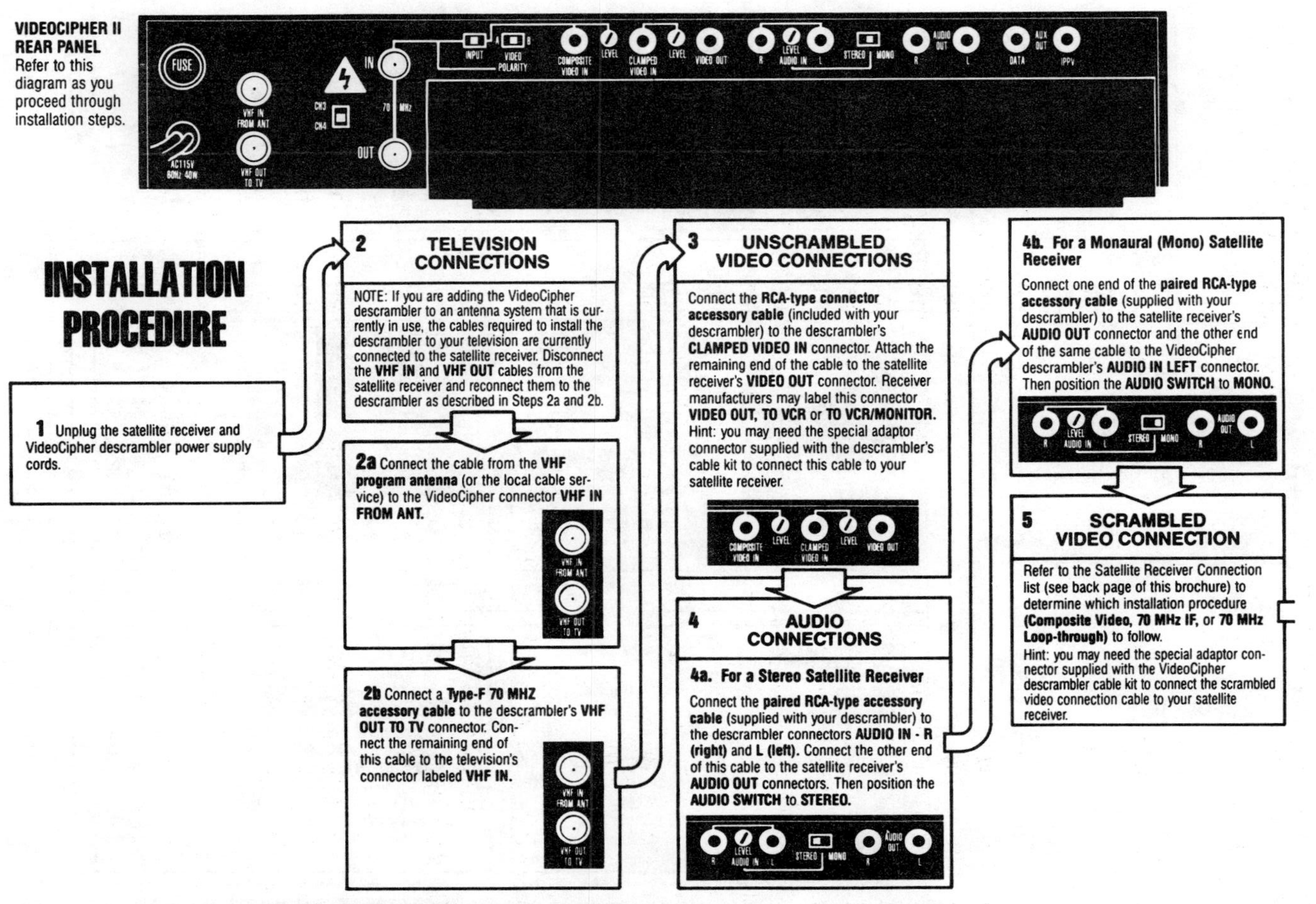

Installation Procedure Flow Chart. *(Courtesy of General Instruments).*

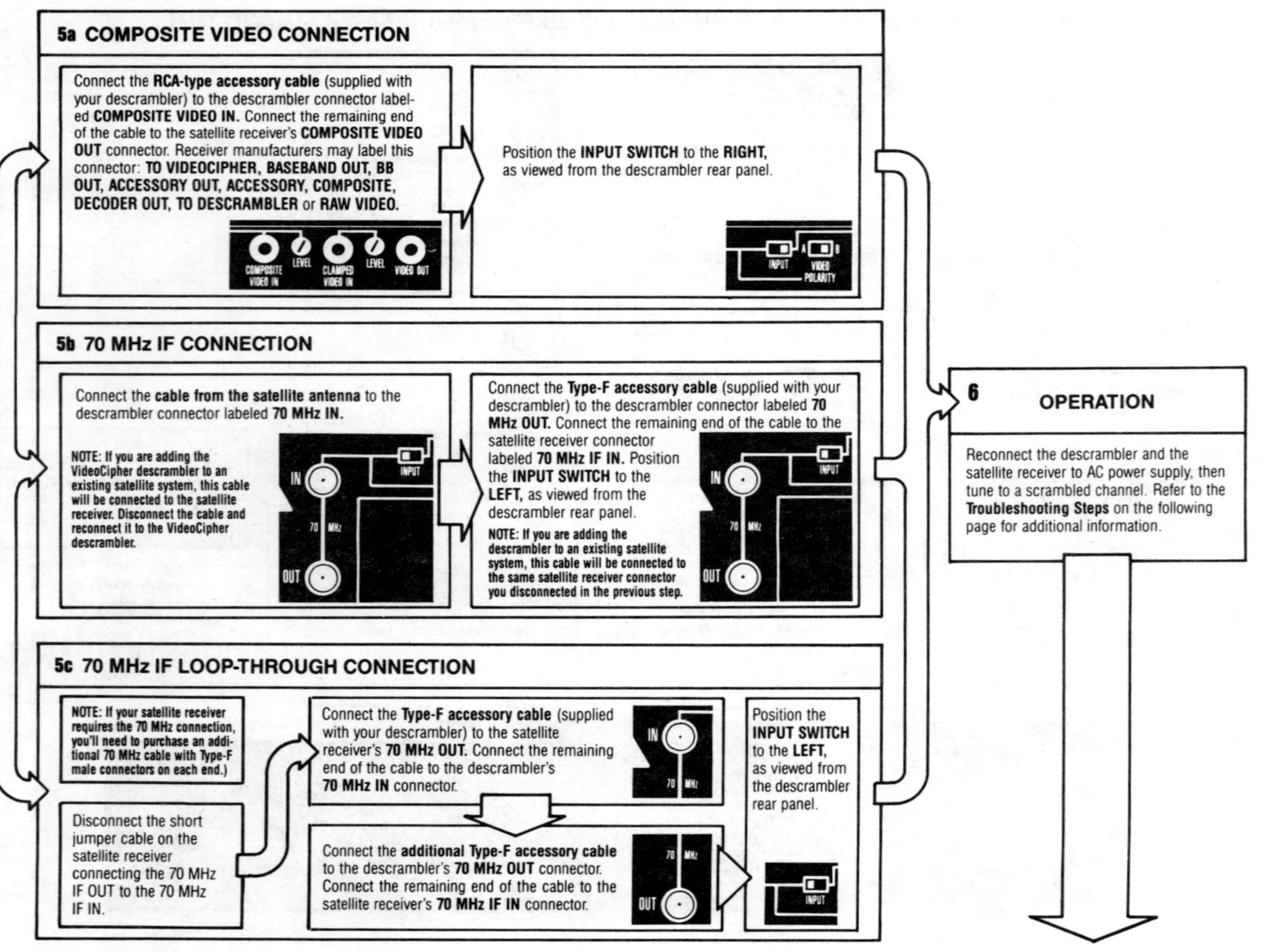
5a COMPOSITE VIDEO CONNECTION
Connect the RCA-type accessory cable (supplied with your descrambler) to the descrambler connector labeled COMPOSITE VIDEO IN. Connect the remaining end of the cable to the satellite receiver's COMPOSITE VIDEO OUT connector. Receiver manufacturers may label this connector: TO VIDEOCIPHER, BASEBAND OUT, BB OUT, ACCESSORY OUT, ACCESSORY, COMPOSITE, DECODER OUT, TO DESCRAMBLER or RAW VIDEO.
COMPOSITE VIDEO IN
LEVEL
CLAMPED VIDEO IN
LEVEL
VIDEO OUT
Position the INPUT SWITCH to the RIGHT, as viewed from the descrambler rear panel.
INPUT
A
B
VIDEO POLARITY
5b 70 MHz IF CONNECTION
Connect the cable from the satellite antenna to the descrambler connector labeled 70 MHz IN.
NOTE: If you are adding the VideoCipher descrambler to an existing satellite system, this cable will be connected to the satellite receiver. Disconnect the cable and reconnect it to the VideoCipher descrambler.
IN
70 MHz
OUT
INPUT
Connect the Type-F accessory cable (supplied with your descrambler) to the descrambler connector labeled 70 MHz OUT. Connect the remaining end of the cable to the satellite receiver connector labeled 70 MHz IF IN. Position the INPUT SWITCH to the LEFT, as viewed from the descrambler rear panel.
NOTE: If you are adding the descrambler to an existing satellite system, this cable will be connected to the same satellite receiver connector you disconnected in the previous step.
IN
70 MHz
OUT
INPUT
5c 70 MHz IF LOOP-THROUGH CONNECTION
NOTE: If your satellite receiver requires the 70 MHz connection, you'll need to purchase an additional 70 MHz cable with Type-F male connectors on each end.)
Disconnect the short jumper cable on the satellite receiver connecting the 70 MHz IF OUT to the 70 MHz IF IN.
Connect the Type-F accessory cable (supplied with your descrambler) to the satellite receiver's 70 MHz OUT. Connect the remaining end of the cable to the descrambler's 70 MHz IN connector.
IN
70 MHz
Connect the additional Type-F accessory cable to the descrambler's 70 MHz OUT connector. Connect the remaining end of the cable to the satellite receiver's 70 MHz IF IN connector.
70 MHz
OUT
Position the INPUT SWITCH to the LEFT, as viewed from the descrambler rear panel.
INPUT
6 OPERATION
Reconnect the descrambler and the satellite receiver to AC power supply, then tune to a scrambled channel. Refer to the Troubleshooting Steps on the following page for additional information.

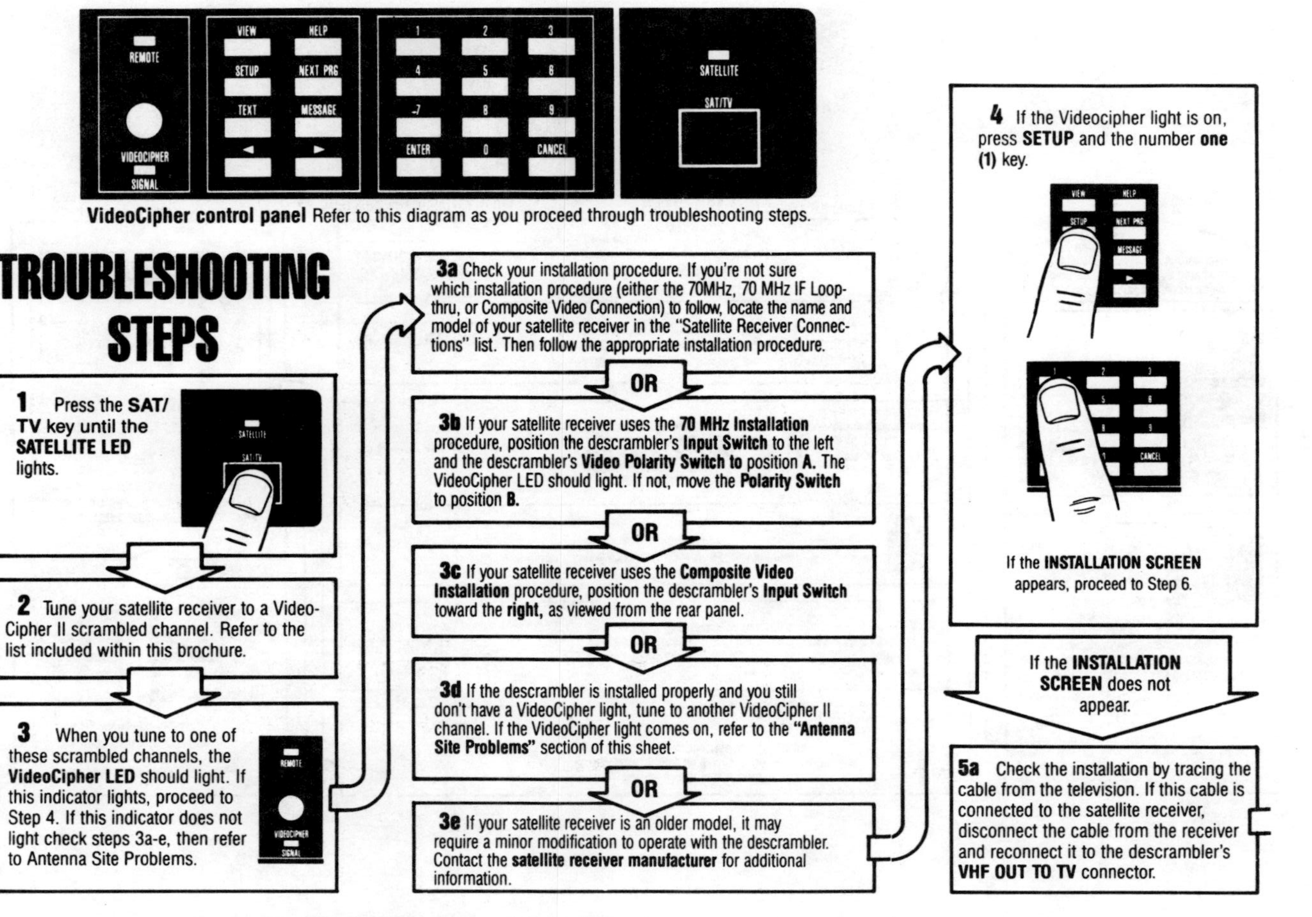

Troubleshooting Flow Chart. *(Courtesy of General Instruments)*

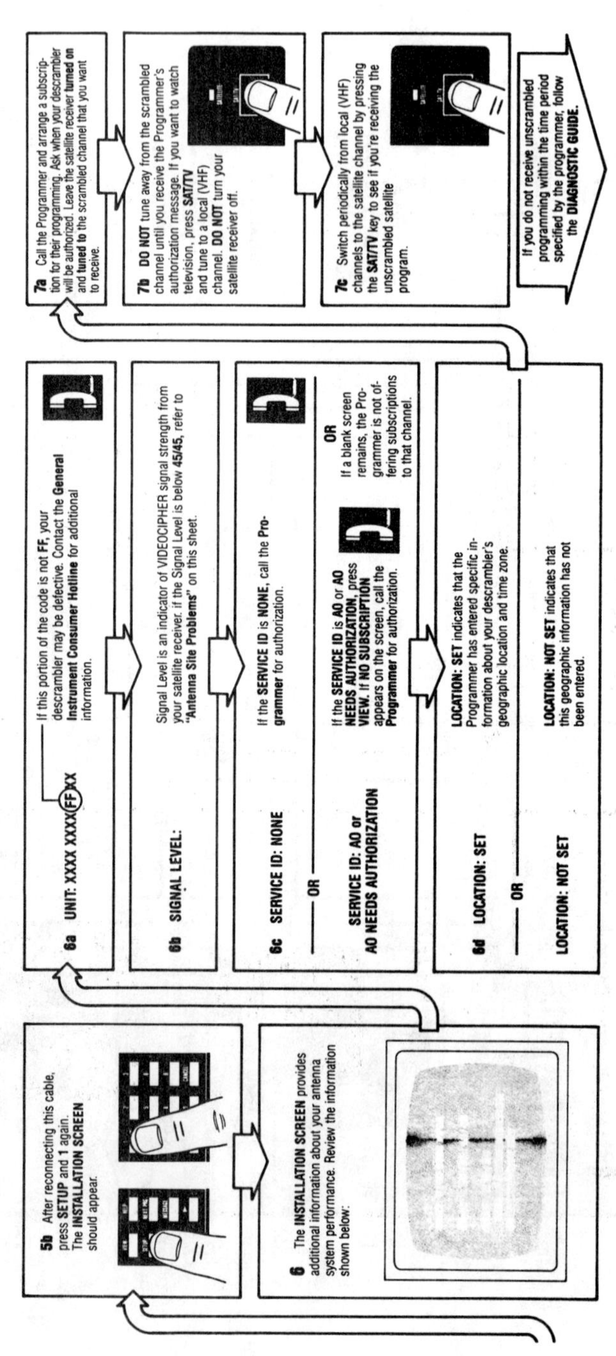

VIDEOCIPHER II DIAGNOSTICS

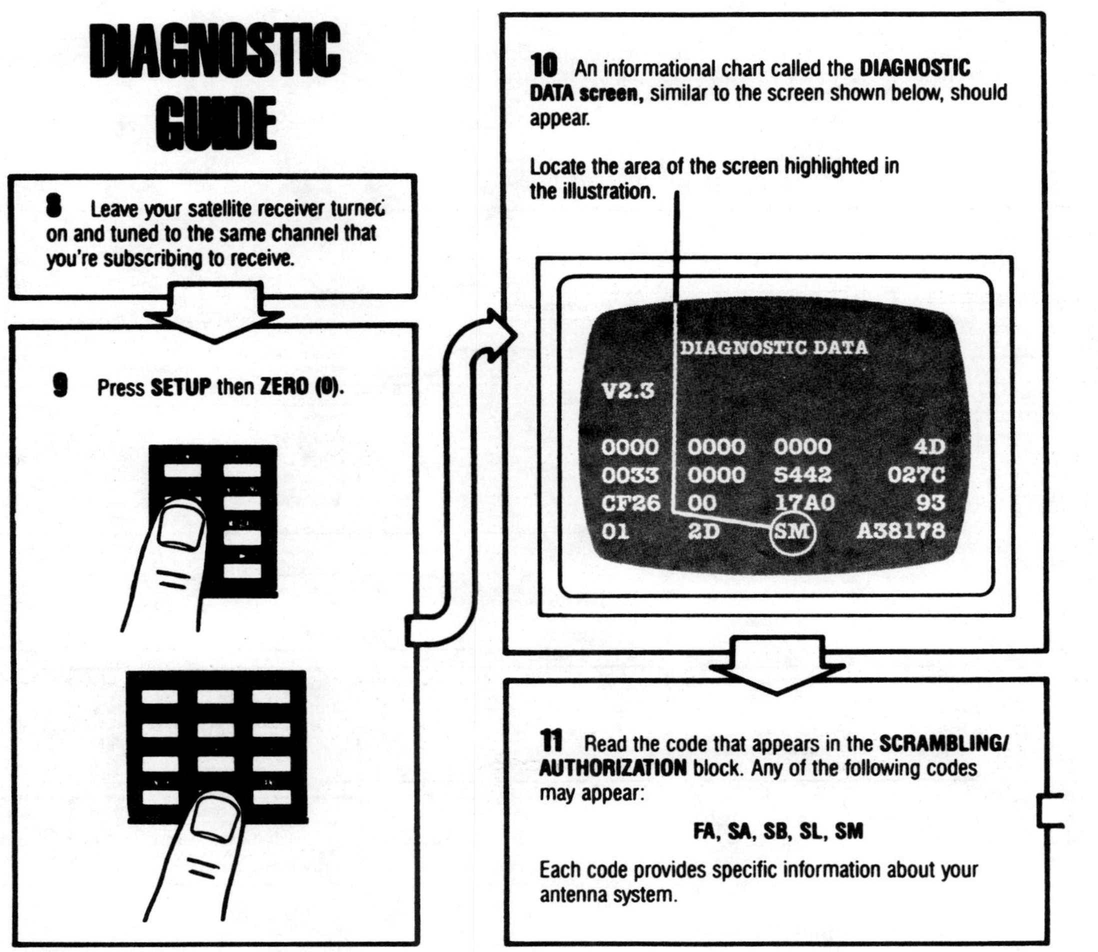
DIAGNOSTIC GUIDE
8 Leave your satellite receiver turned on and tuned to the same channel that you're subscribing to receive.
9 Press SETUP then ZERO (0).
10 An informational chart called the DIAGNOSTIC DATA screen, similar to the screen shown below, should appear.
Locate the area of the screen highlighted in the illustration.
DIAGNOSTIC DATA
V2.3
0000 0000 0000 4D
0033 0000 5442 027C
CF26 00 17A0 93
01 2D SM A38178
11 Read the code that appears in the SCRAMBLING/AUTHORIZATION block. Any of the following codes may appear:
FA, SA, SB, SL, SM
Each code provides specific information about your antenna system.

APPENDIX G

VIDEOCIPHER II DIAGNOSTICS

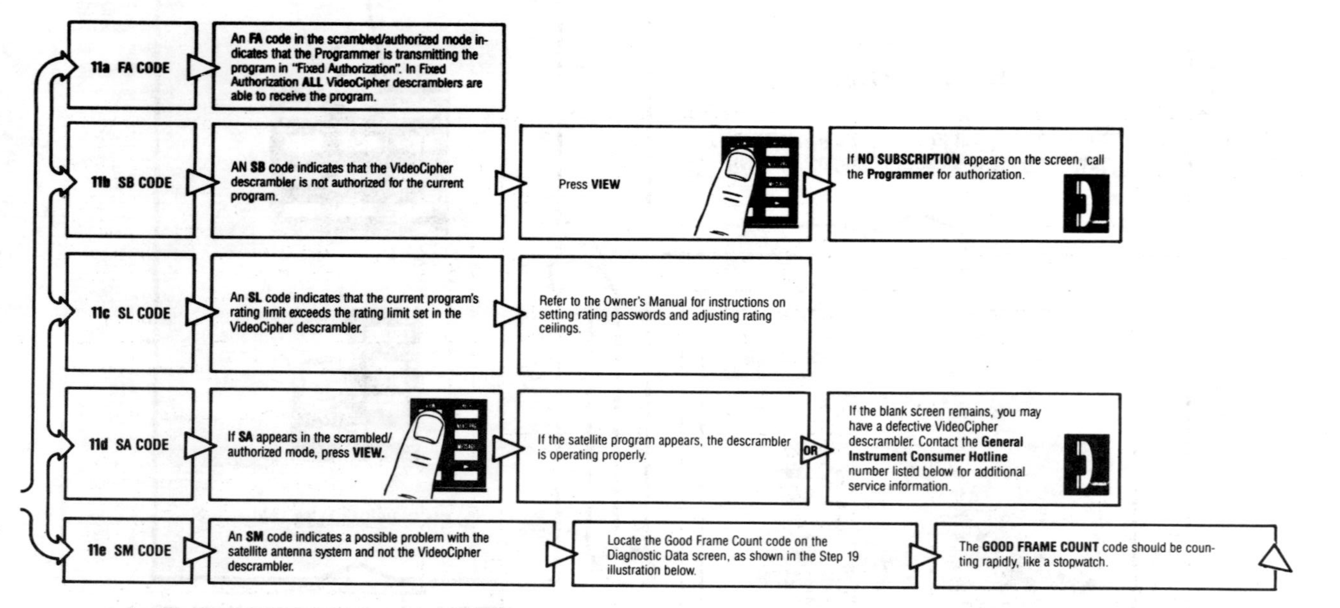

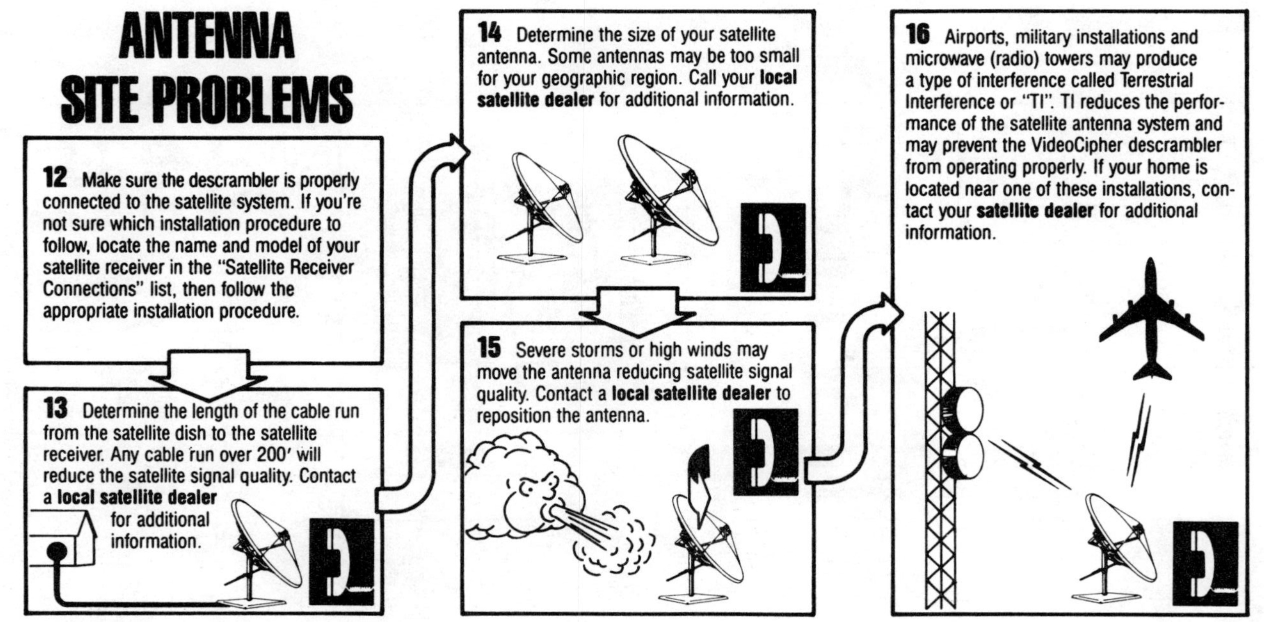
ANTENNA SITE PROBLEMS
12 Make sure the descrambler is properly connected to the satellite system. If you're not sure which installation procedure to follow, locate the name and model of your satellite receiver in the "Satellite Receiver Connections" list, then follow the appropriate installation procedure.
13 Determine the length of the cable run from the satellite dish to the satellite receiver. Any cable run over 200′ will reduce the satellite signal quality. Contact a **local satellite dealer** for additional information.
14 Determine the size of your satellite antenna. Some antennas may be too small for your geographic region. Call your **local satellite dealer** for additional information.
15 Severe storms or high winds may move the antenna reducing satellite signal quality. Contact a **local satellite dealer** to reposition the antenna.
16 Airports, military installations and microwave (radio) towers may produce a type of interference called Terrestrial Interference or "TI". TI reduces the performance of the satellite antenna system and may prevent the VideoCipher descrambler from operating properly. If your home is located near one of these installations, contact your **satellite dealer** for additional information.

VIDEOCIPHER II DIAGNOSTICS

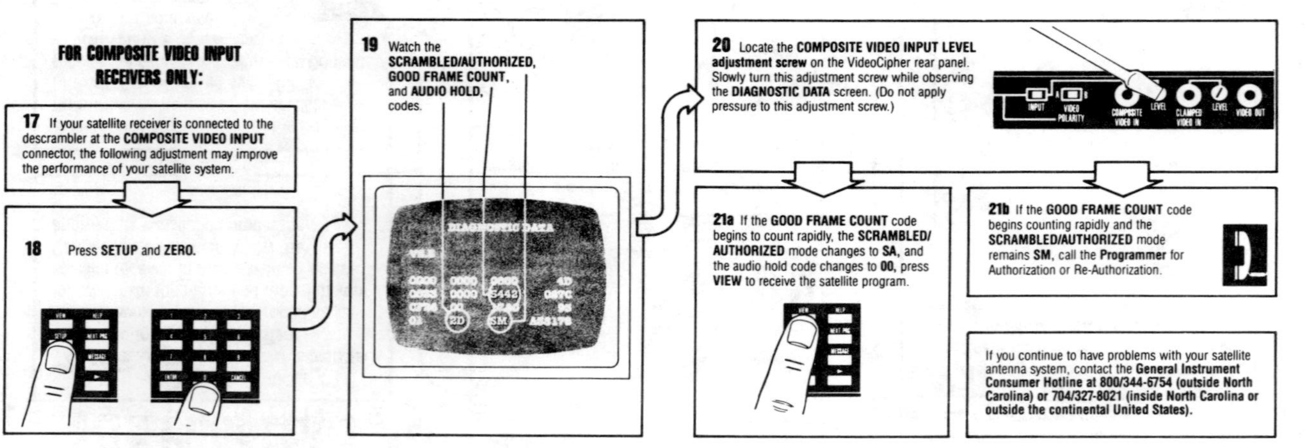

Troubleshooting Diagrams *(Courtesy of General Instruments)*

INTERGRATED RECEIVER DESCRAMBLER (IRD)

In late 1986, several consumer-oriented manufacturers began offering dual-band receiver products which incorporated the VideoCipher II into their overall design. There are several reasons for placing the descrambler module inside the receiver. First, for example, a stand alone VideoCipher II descrambler duplicates some of the audio and video processing functions that are normally handled by the satellite receiver. The on-screen graphics display provided by some receivers may be eliminated, because these features are incorporated into the VC II module giving the user information like satellite being viewed, programmer, movie title, and length of the movie.

By building the descrambler into the receiver, manufacturers are also able to pass along cost savings—up to 50% off the manufacturer's cost for complete descramblers to the consumer.

There is no reason to duplicate two-power supplies, two demodulators, and two audio and video processor boards with a stand alone VideoCipher II decoder and separate satellite receiver. These can be incorporated into one unit.

Installation is also greatly simplified since the receiver and the descrambler are permanently connected. The internal descrambler allows easier operation and the return to full featured remote-control which is lost, when a videocipher II stand alone decoder is installed on a remote control receiver.

The technical specifications for the receiver portion of an IRD are much more difficult to achieve than the requirements of a receiver used with a stand alone videocipher II. The Videocipher II descrambler circuitry is provided to the IRD manufacaturer in the form of a circuit board encased in a plastic cage 9″ long by 11.5″ wide and 1.77″ thick. This module is the heart of all VideoCipher consumer units sold to date. The physical size of the module has caused many headaches for receiver designers, who must find a place for the module in the limited space left in today's high tech satellite receiver designs. This is normally found underneath the audio/video processing board in a special housing with a card edge connector that interface the two. The sealed cage provides an effective means of detecting circuit tampering, as well as ensuring adequate air space for proper cooling. Many IRD designs due seem to run hot, causing instability in the decoder.

Most IRD receivers include large power supplies for internal actuator control. This causes additional unwanted heat. The addition of the descrambler module further increases the weight of the unit by two pounds and with the epoxy sometimes four pounds.

Before an IRD can be sold in the market place with an internal descrambler it must pass a series of very difficult electrical test designed to ensure proper reception of both scrambled and unscrambled video signals without the units microprocessor driven software interfering with the videocipher authorization data.

A typical IRD consists of the VCII module and the receiver unit which contains a FM demodulator, relays, power supplies, a VHF modulator, an audio switch, a keyboard, a microprocessor and other circuitries. (see figure G-3).

The FM demodulator provides scrambled or unscrambled signals to the VCII module after a de-emphasis network and a lowpass filter. For scrambled signals, the VC sync signal from the VCII module controls the relays to switch to the output of the VCII module. The descrambled video signal from the VCII module is fed into a VHF modulator via a low pass filter and a clamper together with the descrambler monaural audio via an audio volume controller.

If the signal is non-scrambled, the VC sync signal controls the relays to switch to the output of the FM demodulator, and the output of the audio. These video and audio signals are fed into the VHF modulator.

The audio switch provides the necessary audio interface and switching between the stereo and monaural audio outputs of the VCII module and the monaural audio output of the audio demodulator. The keyboard and the microprocessor provide the interface between the user and the VCII module to support various consumer features such as Impulse Pay-Per-View, Program Rating Lockout, and Text Services using the On-Screen Display. They also provide the interface between the user and the receiver unit to support channel selection, volume control and muting.

The power requirements for the module are unusual for satellite equipment, but not unusual for computing equipment, which the VideoCipher II is. Dual-polarity 5– and 12– volt supplies are required, with the +5 volt digital supply demanding 1.10 Amps maximum. To preserve the high signal-to-noise ratios available in a digital audio transmission system such as the VideoCipher II, a separate +5 volt supply for the on-board audio line amplifiers is required.

One of the receivers performance specifications for products licensed to contain internal VideocCipher II descramblers is that when the user presses the power switch to turn the receiver off the receiver is required to continue providing both power and satellite signal to the descrambler module and if the VideoCipher requests it, the receiver must change channels. The reason for this unusual requirement is to keep the authorization signal in the module continuously refreshed or updated. As soon as the receiver is turned "off" the VCII module begins to search for a scrambled channel. Since the same authorization messages are transmitted by all programmers that serve the satellite market, any scrambled satellite service will keep the VCII module refreshed, whether or not the unit is authorized to receive that particular service.

The familiar "VideoCipher Signal" light on the front of a stand alone VCII decoder must also be present on the front of all IRD units. The module provides an active low TTL signal to indicate that a VideoCipher II-scrambled channel is being received. This signal is provided to relays that switch between unscrambled video and audio and the video and audio outputs of the VCII module. Although rare, there are instances where audio subcarrier services not affiliated with the active video signal have remained on a transponder when the video service initiated scrambling. Some examples are the radio stations WFMT chicago, which is on GALAXY 1 transponder 3, and WQXR on Transponder 15 both of these video channels are scrambled. Reception of this "piggyback" service on a single-receiver system equipped with a separate component VideoCipher II was not possible without disconnecting the VideoCipher II to remove it from the signal loop. Many IRD receivers should be equipped with a button to allow the defeat of the automatic audio switching, allowing reception of subcarrier audio services on scrambled channels.

One important and very popular feature with IRD's is the use of on-screen displays which replaces all or part of the picture with text containing helpful user instructions. The VideoCipher II makes substantial use of onscreen displays providing information such as installation information, program titles, program time remaining and subscription information. Manufactures of IRD units must make certain that any on-screen displays provided do not conflict with the displays provided by the descrambler.

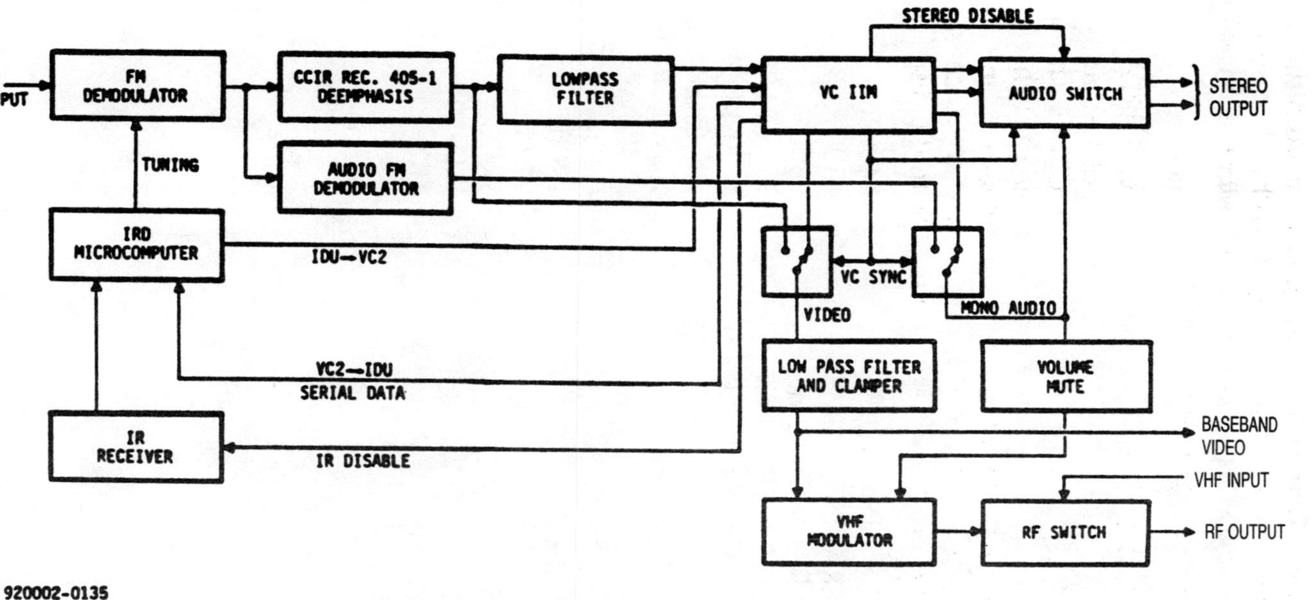

Figure G-3. IRD Block Diagram *(Courtesy of General Instruments)*

Table G-6. IRD Equipment Manufacturers.

MANUFACTURER	MODEL NUMBER
CHANNEL MASTER	6441
CHAPARRAL	CHEYENNE IRD
ECHOSTAR	SRD 3000
ECHOSTAR	SRD 5000
ECHOSTAR	SRD 4000
ECHOSTAR	SRD 7000
ECHOSTAR	SRD 8000
C.ITOH	CITATION 900
C.ITOH	CITATION 1000
GENERAL INSTRUMENT	2400 R
GENERAL INSTRUMENT	2500 R
GENERAL INSTRUMENT	2600 R
HOUSTON TRACKER	TRACKER V
HOUSTON TRACKER	TRACKER VI
HOUSTON TRACKER	TRACKER VII
HOUSTON TRACKER	TRACKER VIII
NORSAT	JR 300 AF
PANASONIC	CRD-4400R
R.L. DRAKE	2400
R.L. DRAKE	2024
STS	SR 100
TEE-COM	IRD-10
TOSHIBA	TRX-100
UNIDEN	7700
UNIDEN	9900